Luuk Simons

Multi-channel Service Design

Luuk Simons

Multi-channel Service Design

Deploying Internet along Other Contact Channels

VDM Verlag Dr. Müller

Impressum/Imprint (nur für Deutschland/ only for Germany)

Bibliografische Information der Deutschen Nationalbibliothek: Die Deutsche Nationalbibliothek verzeichnet diese Publikation in der Deutschen Nationalbibliografie; detaillierte bibliografische Daten sind im Internet über http://dnb.d-nb.de abrufbar.

Alle in diesem Buch genannten Marken und Produktnamen unterliegen warenzeichen-, marken- oder patentrechtlichem Schutz bzw. sind Warenzeichen oder eingetragene Warenzeichen der jeweiligen Inhaber. Die Wiedergabe von Marken, Produktnamen, Gebrauchsnamen, Handelsnamen, Warenbezeichnungen u.s.w. in diesem Werk berechtigt auch ohne besondere Kennzeichnung nicht zu der Annahme, dass solche Namen im Sinne der Warenzeichen- und Markenschutzgesetzgebung als frei zu betrachten wären und daher von jedermann benutzt werden dürften.

Coverbild: www.purestockx.com

Verlag: VDM Verlag Dr. Müller Aktiengesellschaft & Co. KG
Dudweiler Landstr. 99, 66123 Saarbrücken, Deutschland
Telefon +49 681 9100-698, Telefax +49 681 9100-988, Email: info@vdm-verlag.de
Zugl.: Delft, Delft University of Technology, Diss., 2006

Herstellung in Deutschland:
Schaltungsdienst Lange o.H.G., Berlin
Books on Demand GmbH, Norderstedt
Reha GmbH, Saarbrücken
Amazon Distribution GmbH, Leipzig
ISBN: 978-3-639-10684-8

Imprint (only for USA, GB)

Bibliographic information published by the Deutsche Nationalbibliothek: The Deutsche Nationalbibliothek lists this publication in the Deutsche Nationalbibliografie; detailed bibliographic data are available in the Internet at http://dnb.d-nb.de.

Any brand names and product names mentioned in this book are subject to trademark, brand or patent protection and are trademarks or registered trademarks of their respective holders. The use of brand names, product names, common names, trade names, product descriptions etc. even without a particular marking in this works is in no way to be construed to mean that such names may be regarded as unrestricted in respect of trademark and brand protection legislation and could thus be used by anyone.

Cover image: www.purestockx.com

Publisher:
VDM Verlag Dr. Müller Aktiengesellschaft & Co. KG
Dudweiler Landstr. 99, 66123 Saarbrücken, Germany
Phone +49 681 9100-698, Fax +49 681 9100-988, Email: info@vdm-publishing.com

Copyright © 2008 by the author and VDM Verlag Dr. Müller Aktiengesellschaft & Co. KG and licensors
All rights reserved. Saarbrücken 2008

Printed in the U.S.A.
Printed in the U.K. by (see last page)
ISBN: 978-3-639-10684-8

For Angelien, Lievijn and Quirien

And for my parents and grandparents

O Lord, please help me find
Patience to endure what I cannot change
Strength to change what I can
And wisdom to see the difference

(Ancient quick prayer)

Contents

Preface

In two ways this work can be seen as the product of perseverance. First, it is about the Internet, which was declared dead in 2001. Fortunately, the right people continued to do the right things - unlike those crazy cowboys that were simply burning money around the turn of the century. And now the Internet is so much 'alive and kicking' that there is recently a serious buzz emerging around 'Web 2.0'.

The second feat of perseverance is related to my own biography. The start of this research was made in 1999, when I worked for KPN Research. Good progression was made in the first half of 2000, but then things came to a halt until October 2001, due to financial trouble in the telecom sector. That October, Rene Wagenaar took me on at the ICT section of the Faculty of Technology, Policy and Management of the Delft University of Technology for two days per week to continue my research. (Rene, thank you for doing so, and I am still grateful for the urge you put on the 'at least two days per week!') By December 2004, my data collection was finished, and by May 2005 I handed in my manuscript in first draft form, feeling that I was getting close to completion. Little did I know about the textual 'polishing and revising' to come! Over the past 10 years, I have been a consultant, and I have learned that the final 20% towards perfection often represents 80% of the costs. But academics tend to be concerned more with perfection than with costs ☺. Furthermore, after September 2005 I was only at TU Delft one day per week, which definitely slowed down progression. Hence it took me until March 2006 before I had convinced Rene and co-promoter Harry Bouwman to provide their consent to send my manuscript to the committee (at least 'if I promised to change XYZ and PQR'). Luctor et emergo (I found out that I am not such a patient person after all...)!

I am grateful that my colleagues at TNO ICT have been patient with me. In the past two years, there have been several blocks of two to three weeks that I 'sneaked away' to the university to be able to make serious progression. Eeverybody has been most understanding: thanks!

Regarding my professional development I am grateful to all my colleagues and customers. The combination of theory, 'the hard reality' and discussions with the people around me with regard to making sense of the world around us has been very valuable to me. And if it is up to me, I intend to maintain that combination for quite a while in the coming future. Rene and Harry, thank you for the academic growth you have fostered. (Harry, you are of course entirely nerdy when it comes to methodology, but I must admit it was very educational: thanks!) I am grateful to KPN (as an employer as well as a highly valued customer) and more recently my managers at TNO ICT for supporting my development as a professional. I have learnt how to be a consultant, project leader, a coach, a sales person and I even 'experimented' briefly as ad interim manager (for about four months my poor colleagues became guinea pigs during the maternity leave of our department manager– still, I think we all came out okay). All of these skills have been useful in this past research effort. Even the sales skills proved to be useful when I was looking for participating firms and business professionals for my research.

And Angelien, you warned me back then, when I told you I wanted to do a PhD, for the size of this undertaking and the spare time this would cost. But we made it! (And it wasn't even so bad, was it?) Anyway, thanks for being there! (And of course you're the best tutor academically, professionally and personally – always ready with a word of advise – asked for, and unasked for, but nonetheless much appreciated!) More recently Lievijn has entered the equation. Fortunately, he has no idea yet about PhD's. Still, his comments on the world, and on me, are a good source of inspiration and even learning.

In this thesis, we mainly discuss the academic side of our service definition method, MuCh-QFD. There is, however, also a practical side to it. The usefulness of a method is also illustrated by its practical success. On that note I am happy to report that before the ink of this thesis was dry, MuCh-QFD, as well as a follow-up session for making a visual prototype based on MuCh-QFD results, were deployed already. And the organization for which we used these methods was very happy with the results.

Special thanks is due to KPN Research for their contributions to the PLACE project and the 'pre-incubation' of this research, to the Telematica Institute for the PLACE-, BITA- and ISI-projects, to Delft University of Technology for their intellectual, emotional and financial support over the past 4.5 years, and to TNO ICT for the 'in kind' contributions to this research (for example with regard to using the GroupSystems™ decision support system, to graduation students' time, and to the business setting that increased focus in my research).

Last but not least I would like to thank all business participants that provided their cooperation: during the case studies at the start of my research and during the design method tests at the end: thank you very much for all your cooperation!

Most grateful I am of course for the trust you have all had in me! Since I am an optimist I cannot help but believe in my plans, but it is nicer when you're not the only one.

Luuk Simons
July 31st, 2006
Voorburg

1 Introduction to the Research Problem

Choose a job you love, and you will never have to work a day in your life.
(Confucius)

Case Exhibit 1-1 Wall paper '1-800-Pirates'

In the 1980's and 1990's a new breed of competitors entered the US wall paper retail market. Existing retailers dubbed them '1-800-Pirates', and their emergence can be seen as an early case of ICT-enabled channel disruption. They used phone ordering to push traditional wallpaper retailers out of the US market. They encouraged their customers to select products using the sample books and expertise provided by their local retailer stores, write down the product codes, and then order directly via phone, at a 30% - 50% discount to retail prices (Bowersox, 1992). The existing players made various attempts to stem the tide, for example by developing proprietary product codes that were initially unknown to the mail order retailers. However, it all proved to be to little avail, and the industry did change as a consequence of these low cost entrants.

1.1 Problem description, research subject and definitions

In the second half of the 1990's the world experienced the rise of Internet commerce (sometimes referred to as the 'dotcom hype'). This period was characterized by multiple predictions of disintermediation to occur in traditional marketing channels. Physical retail stores were claimed to be 'doomed' in terms of competitive value. Instead, it was expected that the majority of customers would start buying directly from manufacturers. For example, Gilder (1994) stated that wholesale and retail would be eliminated and only Internet intermediaries would survive. Bill Gates (1995) predicted the 'death of the middleman'. And Prahalad (1998) saw disintermediation as one of the eight trends that would transform the economy. By the end of the 1990's, this world view was paramount and all firms on the major stock exchanges were required to present explicit Internet strategies. If not, they were considered to be laggards, and share prices and market capitalization would drop.

In 1999 we as a research team suspected that purely online retailing might have its limits and that there would be significant value in 'click and mortar' approaches. History has proven our suspicion to be right, since we were all witnesses of the Internet bust in 2000-2001, leading to the bankruptcy of many pure dotcoms

(Cassidy, 2002; Preissl, 2004). We started an explorative case study research[1] among 19 firms that combined the Internet with their existing channels (like call centers, physical stores and personal sales representatives) in order to enrich their sales support capabilities. This was the starting point for the underlying as well as other research highlighting management strategies for click and mortar approaches (Steinfield, Wit, Adelaar, Bruins, Fielt, Hoefsloot, Smit and Bouwman, 2000; Steinfield, Bouwman and Adelaar, 2002). One of the main conclusions of this research was that in many instances 'click and mortar' players could generate more competitive power than pure offline or pure dotcoms approaches.

As a result, our interest shifted to the question as to how to combine click and mortar channels effectively. The answer to this question is not trivial. Venkatesh (1999) distinguished 6 different strategies. For example, should online presence be a copy of physical retail presence ('mirror' strategy)? Or should it be an explicitly separate approach, maybe even under a different brand ('parallel' strategy)? Or should a firm follow for example an 'anti-mirror' strategy, which means that business processes in physical outlets are restructured to match the new opportunities that the Internet offers and make them 'Web-aware'? Even now, these types of questions continue to puzzle firms, and we would like to aid them by providing a method for 'click and mortar' service definition choices.

Now let us turn to the other side of the coin: the customer perspective. In recent years, despite the 'dotcom bust', customers have increasingly become multi-channel shoppers, and find it quite natural to use websites and physical stores as part of the same buying process. Already in 2004, Forrester Research showed that 65% of consumers in the US are cross-channel shoppers that search online and buy offline (Wilson, 2004). Also the other way around is popular: 69% of Americans perform orientation offline before buying online (Smits, 2006). Finally, research conducted by Forrester also suggests that cross-channel customers spend on average 30% more than single-channel customers. Similarly in Netherlands and Europe, trend analyses over the past years have shown that, as far as consumer preferences for certain channels are concerned, physical channels have become less important in the 'orientation', 'buying' and 'new needs' phases, while the so-called 'distance' channels (Internet, e-mail, telephone) have become more important (Schueler, 2003). Also in the Netherlands the online orientation phase increasingly determines what is bought where, even if the actual purchase is often made offline (Planet-Internet, 2004), and shopping is increasingly a multi-channel experience (Schueler, 2003).

Thus, firms have a lot to gain by supporting multi-channel shoppers and their buying processes. Our aim with this research is to help organizations use the Internet as part of their marketing and multi-channel service approach. This means combining a customer orientation (delivering customer value) with a

---

[1] A group of 10 researchers participated in the PLACE project. The acronym PLACE stands for: 'Physical presence and Location Aspects in electronic Commerce Environments'. Involved in this project were Charles Steinfield from Michigan State University, Thomas Adelaar, Arnout Bruins, Erwin Fielt, Alko Smit from Telematica Instituut, Harry Bouwman, Els van de Kar from Delft University of Technology, Ellen de Lange and Mark Staal from KPN Research, and Luuk Simons from KPN Research and Delft University of Technology.

commercial orientation (gaining a competitive position and making money) and a channel orientation (which customer support can be provided via which combination of channels).

We start in 1.1.1 by taking a brief look at an historic example of channel disruption caused by the rise of call center based selling. This also illustrates the practical relevance of ICT based disruption of markets (1.1.2). We then introduce our research objective, research questions and research approach, together with the setup of this thesis (1.2). Finally, we address our research philosophy (1.3) and the theoretical contributions we aim to provide (1.4).

Before we proceed we need to explain some of the key concepts we use in our research: channels, e-services or Internet services, and multi-channel mix. To start with the first concept: when we talk about channels, we do so from a marketing perspective, using the following definition: sets of (independent) organizations involved in the process of making a product or service available for consumption or use (Stern, El-Ansary et al., 1996). In this study, we focus mainly on the Internet, call centers, physical stores and the personal sales channel. One of the main functions of sales channels is to support customers throughout the sales cycle. Hence, channels provide services that support orientation, ordering or after sales needs of customers. Which brings us to the second core concept of our research. For us, e-services or 'Internet services' are synonymous and have a service process meaning: 'the process of offering any good or service over the Internet is a service process' (Grönroos, Heinonen, Isoniemi and Lindholm, 2000). By e-services we mean channel service processes whereby customers are engaged via the Internet. The e-services we focus on are 'auxiliary' services (Grönroos, 2000): they are not a core product or service that customers pay for, but 'peripheral' services that aid people in the buying and/or consumption process (e.g. the way questions are answered or information is provided, service recovery procedures, directions for consumption of the core offer, after sales support, etc) (Grönroos, Heinonen et al., 2000; Normann, 2000). Thirdly, when we talk about a multi-channel mix we refer to the total set of marketing channels offered by a supplier. And when we talk about a multi-channel service mix we refer to the total set of 'auxiliary' services they collectively provide.

1.1.1 Historical illustration of the disruptive effects of ICT on existing channels

In this section we discuss a historical example of disruptive ICT for existing channel arrangements. The ICT-based disruption in this example was the introduction of call centers. Historically, channel marketing and distribution took place via one relatively homogenous channel system, based on wholesale and retail stores. Although many different types of players (e.g. public storage warehouses, retail/wholesale distributors, individual stores, retail chains and buyer's distribution channels) could be active in the same market, they roughly based their approach on the same economics: efficiency in logistics and effectiveness in product availability and retailing. This meant making a focused yet sufficiently broad product portfolio available to customers (e.g. furniture, groceries, clothing, etc), offering choice and sales support at a good price. Many channel tasks were performed by physical retail stores in some form, with logistics and warehousing in the background. Examples of those tasks are: supporting information gathering by customers, product sampling, ordering, delivery/transferring physical possession and financial transactions.

In the 1980's these traditional channel arrangements came to be threatened by the emergence of call centers. It is interesting to consider the historical case of mail order companies selling wallpaper, called '1-800-Pirates' by their competitors, as an early case of ICT-enabled channel disruption (see Case Exhibit 1-1). These 1-800-Pirates used phone ordering to push traditional wallpaper retailers out of the US market in the 1980's and 1990's. Their low cost competition based on call centers seriously disrupted the existing market structures (Bowersox and Bixby-Cooper, 1992). This example illustrates how new ICT-based channels can change the added value that customers perceive from existing channels, how they can impact the competitive position of existing players and how they can even change the structure of a market. The emergence of call centers created a situation in which customers suddenly had a new and different type of channel at their disposal. In this respect, adding the Internet to the channel mix is 'just' a next step in the evolution of channel arrangements.

1.1.2   Relevance of Internet for current channel arrangements

There are some remarkable similarities between this historical case and recent Internet issues, although there also some differences. On the one hand, this case illustrates competitive advantages of direct ordering. On the other hand, it shows some of the limitations of phone ordering: since call centers do not offer visual support, customers had to obtain catalogues or visit physical retail stores in order to choose a wall paper pattern they liked. The '1-800-Pirates' type of mail ordering was based on the ability to cannibalize retailers, could afford to offer less service than the physical retailers, and based its competitive advantage purely on price competition. Recent Internet examples like E*Trade and Schwab, on the other hand, indicate that service levels can also increase when the Internet channel is used (see Case Exhibit 1-2). Besides, the combined potential of low cost and high service is potentially more threatening to existing players. The exhibit illustrates that within a decade the nature of competition and the market structure for private investments in the US have significantly changed due to the rise of the Internet.

One of the main threats posed by the Internet to existing channels is based on the fact that it follows 'information economics' logic, thanks to the automation of information exchanges and transactions. In 'information economics', the costs of reproducing information goods are minimal. This implies that exploiting competitive advantages will be based largely on using network effects and scale advantages. After initial investments and production costs are recovered, large profits can accumulate from additional sales (see also Shapiro & Varian, 1998). This implies that many of the commerce activities mentioned previously (information gathering by customers, product sampling, ordering and financial transacting) are often performed more economically by e-tailing than by retailing. Nevertheless, physical stores continue to offer product availability and service advantages, and sometimes logistical advantages. For example, it is cheaper for supermarkets when customers come to the stores (which also serve as local

16

inventory storage points) and 'pick, pack and ship' their purchases via self-service, than when they order online and want home delivery (Simons, Bouwman and Steinfield, 2002).

Case Exhibit 1-2 Disruption in US private investment market: E*TRADE, Schwab, Ameritrade

The cases of E*TRADE and Schwab illustrate significant changes in the market structure for US private investments (Modahl, 2000), due to the extensive online information and transaction services that became available at low costs. In the beginning of 1996, 85% of the market consisted of high end services and 15% were discounters deploying cheap telephone ordering services. In effect a new mid-market segment arose. In 1999 this segment was expected to capture around 48% of the US online investment market by the end of 2003, managing about $1,5 trillion in assets (Modahl, 2000). Although we did not conduct the same study for 2003 or 2004, but we did find several analyses that appear to confirm the numbers. In the second half of 2003, important players like Ameritrade and E*Trade each have over 10% market share of online accounts (Dukcevich, 2004), and market leader Schwab has an even bigger share. In other words, the top 3 players alone cover more than 30% of the market. Moreover, Schwab alone was approaching $1 trillion in managed assets in the first half of 2004 (Slaughter, 2004). This means that the growth in online managed assets has exceeded the 1999 predictions. Finally, the 'traditional discounters', who were expected to have 7% market share by the end of 2003 (Modahl, 2000), recently seem to have disappeared as a separate market segment. The mid-market players have become so cheap and offer such good online service support, that the low-end players have been forced to compete with the mid-market segment to remain attractive for customers (Change-Sciences-Group, 2004). Thus, the nature of competition has changed with the arrival of the Internet.

The '1-800-pirates' case also illustrates the interdependence between channels. The relative value of a channel in a market depends on the other channels that are available. This can also be illustrated via one of the book store cases we encountered in our research: when a national book chain is present at busy locations in towns and at train stations, it may want to include 'in store pick-up and payment' as an Internet function. In this particular case this function (which would not have suited Amazon.com or Bol.nl) proved to be very popular and provided a win-win situation for the channels of this firm. It drew more customers to the physical stores, and as a result revenues associated with impulse buying increased. And their traditional 'in store' customers were stimulated to visit the site by the combination of the brand of the chain, the cooperation and promotion of the store owners and the fact that much more titles could be offered online. Because of this interdependence between channels we want to look specifically at the multi-channel context in our attempt to define new Internet services that have a competitive value.

Finally, the dotcom boom and bust era's both showed many examples of firms that did not know how to respond to the new 'Internet enriched' competitive field: many e-service initiatives were unsuccessful (Cassidy, 2002; Preissl, 2004).

Moreover, they also lacked methods to help them find out which e-services to develop and which multi-channel choices to make. And since Internet sites are by no means the final ICT channel innovations, such methods can also be expected to have value for deploying future innovations. Hence we wanted to develop a method that helps firms define new multi-channel opportunities in response to ICT channel innovations. Though we expect that our method is useful for multiple ICT channel innovations (see also chapter 8), we focused on e-services in a multi-channel context in our research. This focus had the important advantage of ample availability of empirical material. In the next section we describe the corresponding research objective and research questions.

The challenge for firms to decide which e-services to develop can be characterized as a service definition problem. Hence our research focuses on the first tasks within the overall design process, in the design phase we identify below as the 'initiation phase'.

1.2   Research Objective, Research Questions and Approach

Our *research objective* is to aid organizations in designing 'auxiliary' e-services (Grönroos, 2000) that have to function next to existing channel services (via for example call centers, physical retail stores, sales representatives and possibly other existing e-services). Currently, there are no methods available to support organizations in this task. Thus, we have developed and tested an e-service definition method that is tailored specifically for e-services in a multi-channel context.

Chronologically, we took the following steps in our research. First we conducted an explorative case study, investigating 19 businesses in the Netherlands that implemented e-services along existing channels (see chapters 2 and 3). Based on this research we decided that e-service definition decisions in the initiation phase of new e-services needed improvement, and that we wanted to develop a method specifically for that phase. (Improvement is meant in terms of incorporating important aspects like customer value or multi-channel service cohesion, for example, and in terms of process characteristics, for instance design process focus. What the exact meaning of 'improving the e-service definition process' should be, was part of our research and is further described in chapter 3.) We identified requirements for such a method and we evaluated several existing methods in relation to those requirements (see section 3.3). The conclusion in terms of design knowledge was that a modified version of the QFD (Quality Function Deployment) approach would likely have the highest performance in relation to our design requirements. Two important considerations for modifying QFD were: speed ('ready in half a day') and incorporating multi-channel issues.

Based on this evaluation of existing methods, we developed our own method: 'Multi-Channel QFD' (MuCh-QFD), see chapter 4. This method was pre-tested and refined in three pre-test rounds, see section 5.4. The final format consists of a standardized intake and a service definition session of four hours with multiple stakeholders. The MuCh-QFD method is described in section 4.2. We also wanted to test the effectiveness of this format in a rigorous way. Hence we chose a research design similar to 'static group comparison' (Hagenaars and Segers, 1980), which we called our 'structured field experiment'. In this research design, the effects of our MuCh-QFD session are compared with effects of a control group session (see chapter 6 for our research design and methodology). We developed a control group session based on an alternative design approach and called it the 'Fundamental Engineering' (FE) session, see section 5.3. We indicated what the expected contributions of MuCh-QFD sessions were in relation to the design requirements in Table 5-3, and what those of FE sessions were in Table 5-4. We also developed formal measurement instruments, besides collecting qualitative data from interviews, questionnaires and group sessions (see chapter 6). Finally, we conducted our field experiment, analyzed whether the sessions had the effects we expected and discussed (dis)advantages of MuCh-QFD and FE under various conditions (see chapters 7 and 8).

In the remainder of this thesis we describe our research on the basis of several research questions. Our overall *research question* is:

How to design e-services in a multi-channel context?

Because this is still a relatively abstract question, we break it down into several sub-questions. The questions we address are:
1. What determines the competitive value of a channel within a multi-channel mix, and where can opportunities for the Internet be expected? *(This question is addressed in chapter 2.)*
2. How can design methods aid the process of designing new e-services that have to function in a multi-channel context, and what are the requirements for such a design method? *(This question is addressed in chapter 3.)*
3. How can we develop an e-service definition method that meets the requirements mentioned in the previous question? *(This question is addressed in chapter 4.)*
4. Which e-service definition method can we use as a control group condition? *(This question is addressed in chapter 5.)*
5. How can we evaluate the performance of our (experimental and control group) e-service definition methods with regard to the design support requirements? *(This question is addressed in chapter 6.)*
6. How do the e-service definition methods perform with regard to the design support requirements? *(This question is addressed in chapter 7.)*

We now discuss the details of our research. Our approach consists of two phases: an exploration phase and a test phase, as illustrated in Figure 1-1. The exploration phase addresses the first two sub-questions, concerning the strengths of channels (question 1 / chapter 2) and concerning design processes and design support methods (question 2 / chapter 3). In Figure 1-1, they are presented side by side, because they can be seen as parallel issues. An e-service design team faces a 'what' and a 'how' question, which are addressed separately in chapters 2 and 3. In chapter 2 we research what the Internet can potentially add to the other channels. However, design teams also face important how questions regarding the design process: how to choose between Internet functionalities, how to match Internet solutions to customer and business needs, how to position Internet services next to other channels like retail stores or call centers? In chapter 3 we focus on the design process and look at ways in which design support methods can support that process. Since chapters 2 and 3 address issues in parallel, they also have parallel theoretical and empirical sections. Chapter 2 addresses theory on channels, channel competition, service elements of channels and on services marketing concepts. By contrast, chapter 3 mainly addresses theory on service development approaches within services marketing literature and on design support methods. As an empirical basis for chapters 2 and 3, we used an explorative case study across 19 cases in the Netherlands, although in each chapter we focus on different aspects of the case information we collected. In chapter 2 we focus on how channels contribute to the overall channel mix. In chapter 3 we focus on the process of how new Internet services have been positioned, designed and implemented in the cases we studied.

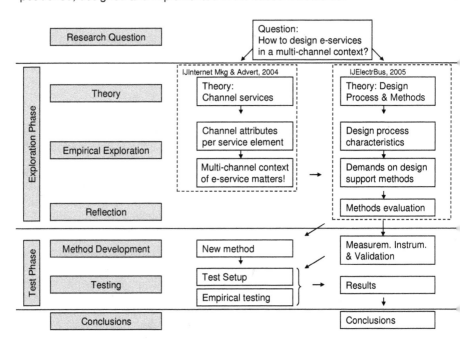

Figure 1-1 Overview of research approach

The main link between chapters 2 and 3 is of an indirect nature: an important conclusion from chapter 2, summarized in the figure as 'multi-channel context of e-service matters', states that the added value of an e-service for customers is always relative, depending on which other channel services are present. This means that when developing a new e-service it is prudent to optimize its services in relation to the other, existing channels, which constitutes part of the design challenge in chapter 3.

In the second phase of our research, the test phase, we take several steps. First, in answer to research question 3 (chapter 4), we develop a new e-service definition support method. Alter (1999) distinguishes three phases in the design process: initiation, development and implementation. We focused on the initiation phase in the test phase of our research. Thus we addressed objectives, scope and high-level functional specifications for the e-services. We chose this focus for two reasons. Firstly, our case studies showed that a systematic service definition approach could be useful, especially in the initiation phase. And secondly, we wanted to test the contributions of our design support suggestions. By limiting ourselves to the initiation phase (which takes less time than an entire design process) we were able to compare the effects of our method across six cases, which received identical e-service definition support.

To answer question 4 - what to take as control group e-service definition method - we look at various options and develop a control group method in chapter 5. Our research design is described in chapter 6, including the development and validation of our measurement instrument. We defined our measurement instrument on the basis of the design requirements presented in chapter 3. To summarize our research design: we formed two equivalent design teams for each of the six cases, whereby one team used a control group method and the other our new, experimental method. This quasi-experimental design creates the opportunity to rigorously test the effectiveness of our methods, even though we were unable to control for all case-specific aspects. With regard to these uncontrolled elements, and to capture a broad picture, we used various data collection methods, for instance independent observations using formal observation protocols and collecting participant feedback. In chapter 7 the results in relation to our design support requirements are evaluated, both qualitatively (via participant feedback and cross case analysis) and quantitatively, based on our measurements. Finally, the conclusions of our research are presented in chapter 8.

1.3   Research Philosophy

In our research we have attempted to combine services marketing approaches with design approaches. This means that we take a design perspective in relation to existing services marketing theory, as described in 1.4. And as far as our research approach and philosophy are concerned, it means that we have combined two different academic traditions. As described below, we followed the 'causal model' tradition of hypotheses formulation and empirical testing, as well as the practical 'engineering' tradition of method development and testing practical effectiveness in real-live situations.

Services marketing research has firm fundaments in experimental research, hypothesizing and testing of causal models (e.g. what are determinants for service quality, for loyalty, for intentions to use an Internet site, etc). Design approaches more generally assume causality and aim at developing innovations that work. The main test for a new design is whether or not it actually does what it is supposed to do.

Within the two traditions differences exist between an 'empirical' and a 'practical' claim, and between an empirical cycle and a design cycle (Aken, 2004). This is illustrated in Figure 1-2. A drawback of the design tradition, with its practical focus, is that a design is often tested by implementing it in one or two cases, after which the results are evaluated and improvements to the design are suggested. This is not the same as rigorous scientific testing. Empirical research on user testing shows that feedback obtained from one or two test cases is still rather case-dependent (one or two test cases generate only 30% and 50% respectively of all potential user feedback, see Appendix D). In our research we want to move one step closer to repeatable results, without losing the value of the design orientation. We do this by testing our method in a more rigorous way: with more cases, using formal measurement instruments and by statistically analyzing results in comparison to control group conditions. On the other hand, a drawback of rigorous testing is that only a limited number of variables can be tested in an experiment, and that there is always a question as to the external validity of the results. In other words: An experimental setting is always artificial, and the experimental findings may have limited validity as far as the real world is concerned.

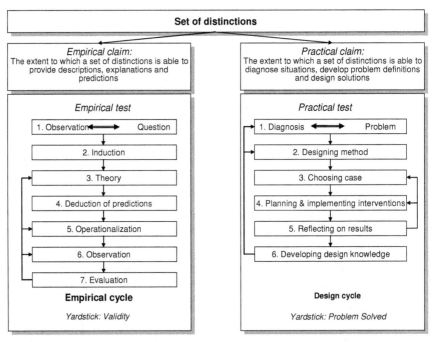

Figure 1-2 The empirical and design cycle, based on van Aken (2004)

We have attempted to develop a synthesis between the two traditions. This has two consequences. Firstly, our method has both a design orientation, i.e. designing a method that helps improve (e)service definition processes, and a causal orientation, i.e. testing hypotheses on the effectiveness of a design method. Secondly, our 'structured field experiment' holds the middle between a strictly controlled experiment and a multiple case study. We develop hypotheses for relationships between dependent and independent variables that we test via a rigorous experimental research design. This is in line with the positivist views on reproducibility of results and using hypotheses that can be tested. However, we will evaluate our results not only through the use of scores on formal measurement instruments (tailored to measure the main experimental effects of our research design). We also collect qualitative case information and perform cross-case analyses. These analyses yield insights and conclusions that are not tested with equal rigor, but that are still useful for further theory development (Yin, 1994). This is in line with the hermeneutic view on the world, which relies on the interpretation and understanding of phenomena in context (Dilthey, 1900; Gadamer, 1960), and is a less strict approach to knowledge creation than positivist experimentation.

As far as the issue of combining design and experimentation is concerned, we have focused more on developing a new service definition method and validating its effectiveness, than on defining and validating a causal model about multi-channel service innovation. In that sense we mostly followed the design cycle. However, during design cycle phase 4 - the planning and implementation of our interventions - we also introduced the second half of an empirical cycle: from theory we deduced predictions about the effects of design tasks and we choose an experimental design with measurement instruments to arrive at a more rigorous evaluation of design task effectiveness. We used the same design approaches for six cases, with 49 participants, and measured the effects for each design task that was performed in comparison to control groups that followed a different method. Nevertheless, each case had unique characteristics (sometimes called 'disturbing factors' in experimental terminology), which provides unique opportunities for gaining additional insights regarding the usefulness of design tasks in relation to case context. So our approach is also similar to case study analyses (Yin, 1994) including use of multiple evaluation instruments in a so-called multi-method approach (using external observers, participant questionnaires, interviews, and debriefing evaluations a month later) to obtain a relatively complete picture. We hope that by following this relatively rigorous research design, we also contribute to the design cycle by developing robust and reproducible knowledge on the effectiveness of the methods we tested.

1.4   Theoretical Contribution

Our research has a relatively unique orientation due to the combination of three central elements. Although each of these elements is discussed separately in literature, little research has as yet been conducted to find out what the consequences are of combining them. *To start with the first central element: we consider providing sales support via channels to be essentially providing services* (e.g. educating customers on the features and benefits of new products and services, helping them between competing offers, providing different payment options, etc). In channel literature it is stated explicitly that different service levels can coexist, as long as they are on different points on a price/quality continuum that customers value and are willing to pay for (Bucklin, 1966, 1972; Stern, El-Ansary and Coughlan, 1996). However, the same literature provides only limited theory on ICT (Information and Communication Technology) or on the strengths and weaknesses of various channels in a multi-channel mix (Bowersox and Bixby-Cooper, 1992; Stern, El-Ansary et al., 1996).

Others start from a transaction cost perspective. The advantages of ICT-enabled channels are generally linked to operational, scale, supply chain disintermediation and cost advantages (Malone, Yates and Benjamin, 1987; Wigand and Benjamin, 1995). This perspective is also dominant in literature that emphasizes electronic channels as such (Negroponte, 1995; Keen and Ballance, 1997; Shapiro and Varian, 1998; Evans and Wurster, 2000). Although this literature does provide several inspiring case examples of ICT benefits, it leaves channel service design issues virtually untouched.

By contrast, the micro-economics of providing service during sales interactions is better described in channel- (Lynn, 2000) and services marketing (Heskett, Sasser et al., 1997; Grönroos, 2000). A disadvantage of channel literature is that the main challenge is said to be marketing positioning and sales effectiveness. What services marketing literature adds is the challenge of finding and orchestrating the right service concept: which mix of services is ideal for our target customers and at the same time suites our own capabilities and provides a lasting competitive edge (Heskett, Sasser and Schlesinger, 1997; Porter, 1999)? Although it is recognized that (information) technology can play a role here (Johnston, 1999; Normann, 2000), theory on where and how to deploy ICT innovations remains mostly untouched. This means that as yet many questions regarding how to use ICT for service mix optimization remain unanswered.

Our second central element is twofold: a) the competitive value of sales support that channels provide is not a static economic fact or objective given. The information offered by the Internet, for example, is not always of a higher quality, nor can it invariably be offered more economically, than the information offered by a call center, even though the Internet certainly offers opportunities in this area. And b) sales support via channels can and should be consciously designed, since there are major opportunities for improvement. In other words: channels are rather versatile and their potential functions can vary significantly depending on the way they are used. In this respect, we agree more with the view that the Internet can deepen a lasting competitive position when it is used strategically (Tapscott, 2001), than with the view that it is like a commodity 'light switch' that all firms can equally 'switch on' to raise the value/cost levels offered to their customers (Porter, 2001).

Hence, our second main element runs counter to the search for a context-independent 'golden standard' which would express 'the best way to deploy the Internet'. Contrary to others, we have decided not to write down 'the most complete list of Internet functions' that can function as an ideal norm for all organizations (Verhagen, Vries and Ham, 2001), or for standardized (e-)service quality standards like SERVQUAL or WebQual that can only be measured post hoc (Parasuraman, Berry and Zeithaml, 1985; Parasuraman, Berry and Zeithaml, 1993; Barnes and Vidgen, 2001). We argue that improving the use of the Internet channel is not the same as maximizing the breadth and depth of online functionality or maximizing WebQual items on all occasions. We would even argue that when it is decided that a requirement like 'clear and understandable product information' is priority number one, at a more detailed level this can mean something very different to a 'nerd' who wants to compare the technical specifications of a product, than it does to a low-tech consumer, who primarily wants to know if he will be able to use the product and how long it will last. In principle, every additional (self)service that is offered to customers has to be justifiable in terms of fitting the target segment and increasing sales per customer, the total number of customers or customer retention at lower or competitive cost levels. This is a design challenge where explicit design support can help.

Our third central element is that the design of channel services can and should be supported by well-structured and -tested design approaches. In services marketing literature it is stated that service quality is crucial to market success, and that design is a key factor in determining quality (Grönroos, 1993; Tax and Stuart, 1997; Johnston, 1999). However, tools for enabling design are relatively scarce in literature on services marketing and management (see also chapter 3). More specifically, there are hardly any tools available in traditional service management literature with regard to designing and redesigning a marketing channel mix for optimal service (Ramaswamy, 1996; Heskett, Sasser et al., 1997; Normann, 2000). In one of his overview articles, Johnston (1999) argues that additional steps in the research on service design and on the ICT-enabling of services are needed. He argues that the most important question in practice is not in knowing what the SERVQUAL dimensions are, but in knowing how to ensure high quality service delivery from day to day, week to week and year to year, based on good service design and service management.

It is in the area of designing Internet channel services (or e-services) that we have focused the theory building and testing efforts of our research. However, although our research contains these testing efforts, in honesty we do not think that our main contributions lie in that area. This would be somewhat overly ambitious, since there hardly is a body of knowledge to be tested in the (multi-channel) e-service definition domain. What we have tested is the effectiveness of certain design tasks and whether the effects of those tasks were in line with expectations from theory. We did this in a quasi-experimental way, closely related to case study based theory generation to create design knowledge, and to draw overall conclusions on theory and implications for management.

Part I: Exploration Phase

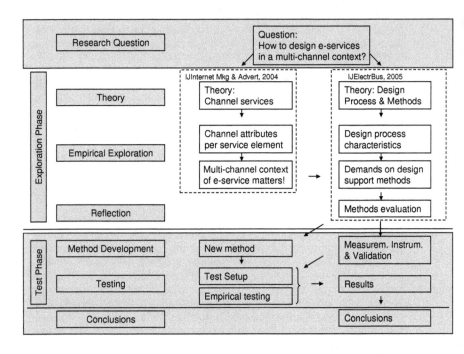

2   The competitive value of channels for service provisioning

An inventor is an engineer who does not take his education too seriously.
(Charles F. Kettering)

Case Exhibit 2-1 Strengths and weaknesses of the Internet and of physical stores

Apart from variations across firms, there are clear overall differences between channels. If we look at the Web, for example, we see that customer benefits are mostly related to pre- and after-sales information and decision support, in combination with 24x7 service access and a sense of control over the (self)-service process This is especially true for information-intensive processes like researching mortgages or technical self-help. In various instances customers have a preference for the 'non-pushy' and 'non-intermediated' nature of self-service. In one case the slower but self-managed searches were explicitly preferred over interpersonal searches (Simons, 2001). The biggest disadvantage of the Web is delayed product possession. For suppliers, low cost self-service opportunities and process efficiencies are the most prominent advantages. Also, although there are some low cost options for lead generation and customer education, the handling of complex leads requires other channels. Interestingly enough, the main issue regarding product complexity is customer perception: we have seen highly complex products being bought routinely over the Internet, because of experience and familiarity with purchases of those types of products on the part of the customers.

The retail store channel, on the other hand, is important for the physical aspects of buying: experiencing the look and feel of products, having immediate product possession, and also offering facilities for product returns and repair. Interpersonal aspects like providing more in-depth decision support and creating assurance and trust are important. Supplier benefits are the stimulation of (additional) purchases through uncertainty reduction, and proximity to customers, which creates a certain lock-in. With regard to fast-moving consumer goods (e.g. groceries) there are supplier advantages as a result of the fact that the store contains the inventory, and because self-service logistics (product pickup by customers) is the most efficient and cheap way to service customers. Supplier disadvantages are the limited possibilities (or high costs) for offering a product range comparable to what e-tailers can offer.

In this chapter we answer our first research question[2]:
1. What determines the competitive value of a channel within a multi-channel mix, and where can opportunities for the Internet be expected?

To answer our research question we take several steps. We begin by discussing channel theory in section 0, where we show that one of the main functions of channels is to support customers throughout the sales cycle. In other words, channels provide services. They do so at a certain price/quality ratio, and opportunities for new channel services lie in finding new price/quality opportunities. Based on theory, we develop a framework of the service elements that are provided during the sales cycle, after which we use explorative case studies (2.2 and 2.3) to compare the added value of several channels (the Internet, call centers, physical retail and personal sales representatives) in terms of this framework. Then we analyze our results (2.4) and present our conclusions (2.5).

We focus on four channels: the Internet, call centers, retail stores and personal sales, and on the ways they complement each other in an overall service offer. These four channels are widely used, are likely to coexist, are often backed up by separate departments (thus qualifying as separate marketing channels) and are relatively different from each other, thus providing a wide array of service facilities. Internet and call centers are made possible by information and communication technology and can be considered as 'distance channels', whereas retail stores and personal sales require physical proximity. It is for this reason that we will refer to our channels as either ICT-enabled or physical channels.

2.1 Theoretical Background

Our literature review is based on service management and marketing channel management research. Service management literature is strong on service design and management. However, with regard to design of a marketing channel mix for optimal service, hardly any modeling takes place in traditional service management literature (Heskett, Sasser et al., 1997; Normann, 2000).

Key concepts borrowed from service literature are the 'total service offer' or 'package', which consists of a 'core product or service' , plus 'peripheral' ,'auxiliary' or 'hidden' services (e.g. the way questions are answered or information is provided, service recovery procedures, directions for consumption of the core offer, etc) (Grönroos, Heinonen et al., 2000; Normann, 2000). 'Auxiliary services' are often non-billable, and they are not primarily what the customer pays for, but they have a large impact on customer satisfaction and the effectiveness of the sales cycle (Grönroos, 2000). Different communication channels (Internet, call, and personal channels as used in stores and personal sales) that market and sell the same core product or service can be said to have different sets of auxiliary services. In literature a distinction has been drawn between channels that provide either 'high support' or 'low support' to customers,

---

[2] This chapter is based on an article published in the International Journal of Internet Marketing and Advertising, Volume 1, Issue 3, 2004.

while selling the same product (Lynn, 2000; Simons, Bouwman et al., 2002). In this chapter, we look at auxiliary services from the following perspective: What type of support and services do they offer the customer around the core product or service? Furthermore, service management literature offers little insight on which core or auxiliary service elements could be supported by which (characteristics of) ICT-enabled or physical channels or combinations of both.

A dominant paradigm in service literature is formed by the SERVQUAL model, which describes service dimensions. SERVQUAL has been used in studies on ICT-systems (Kettinger and Lee, 1994; Pitt, Watson and Kavan, 1995; Jiang, Klein and Carr, 2002; DeLone and McLean, 2003), but rarely in relation to ICT-based-services (Parasuraman, Berry et al., 1985; Parasuraman, Berry et al., 1993). There is a question as to whether the five dimensions of SERVQUAL - tangibles, responsiveness, reliability, assurance and empathy - actually capture the perceptions of ICT-based service quality and whether they take the differences between the various channels into account. In one of his more recent overview articles, Johnston (1999) argues in the research concerning service design and the ICT-enabling of services, additional steps are needed. He argues that the most important question in practice is not in knowing *what* the SERVQUAL dimensions (Parasuraman, Berry et al., 1985) are, but in knowing *how* to ensure high quality service delivery from day to day, week to week and year to year, based on good service design and service management.

Channel literature is strong on marketing channel strategy and structure (e.g. balance of power, channel conflicts, market reach, etc). In literature, some remarks are made on channel service elements (although no overall or generally accepted set of elements is defined), and again very few specifics on the use of ICT in channels are offered. One of the exceptions is a recent paper investigating competitive strategies in click-and-mortar retailing (Steinfield, Bouwman et al., 2002). Other recent contributions to the field focus on Internet-based service quality (Barnes and Vidgen, 2001; Bhattacherjee, 2001; Verhagen, Vries et al., 2001; Chen, Gillenson et al., 2002). However, multi-channel service comparisons are still largely unexplored, as is the matter of design support for a multi-channel mix.

Traditional marketing channel literature offers little in terms of theory on ICT or on the strengths and weaknesses of the various channels in a multi-channel mix (Bowersox and Bixby-Cooper, 1992; Stern, El-Ansary et al., 1996). The advantages of ICT-enabled channels that are mentioned are usually linked to a transaction cost theory perspective, quoting several operational, scale and cost advantages (Malone, Yates et al., 1987; Wigand and Benjamin, 1995). This perspective is also dominant in literature that is more focused on ICT-enabled channels as such (Negroponte, 1995; Keen and Ballance, 1997; Shapiro and Varian, 1998; Evans and Wurster, 2000).

In contrast, traditional channel literature devotes much attention to market forces and to the fact that each channel should be aimed at supporting specific customer needs. Channels operate in a market, and are in the business of connecting producers to buyers, supporting customers during the buying cycle (Lynn, 2000), offering maximum quality at minimal costs. According to channel theory, channel arrangements in the long run evolve towards the 'normative structure' (Bucklin, 1966). In this structure, expensive, high support channels can co-exist with low cost channels that offer limited support, as long as different customers choose different cost-service combinations. In its simplest form, this normative structure could be modeled as a linear price-quality line, with different channels or firms deploying multiple channels, occupying different points on the line. See for example firm X, with a channel set X, and competitors A and B (Figure 2-1). Movements to a higher price-quality line are illustrated by arrows 1 and 2. Movement 1 means that firm X is likely to become a threat to competitor A, by approaching price levels that customers of competitor A are willing to pay, while offering superior quality. Movement 2 means that firm X is likely to become a threat to competitor B, by approaching quality levels that customers of competitor B are accustomed to, but at a much lower price.

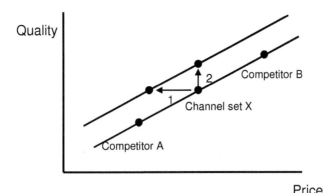

Figure 2-1 Market equilibrium changes when channel set X moves to upper Q/P line

Using the Internet or other forms of ICT allows companies either to reduce the cost of a service, or increase service levels. In some industries (e.g. financial services) the impact has been quite disruptive to existing channel systems and propositions, as can be illustrated with the cases of E*TRADE and (e)Schwab (Modahl, 2000), see Exhibit 1-2. Due to the extensive online information and transaction services that came available at low cost, the market structure for US private investments changed significantly. In the beginning of 1996, 85% of the market consisted of high-end services, and 15% were discounters (deploying cheap telephone ordering services). A new mid-market segment emerged (which was estimated to cover approximately half of US private transaction volume by 2003), with extensive on-line services plus complementary telephone and off-line support, at a price slightly above the deep discounters. The new proposition attracted many customers from the two other market segments. This can be seen as a combination of movements 1 and 2 in Figure 2-1. Recently, it even appears that the mid-market players have become so cheap, and offer such good online service support, that the low-end players have been forced to compete with the mid-market segment to remain attractive for customers (Change-Sciences-Group, 2004). Thus, the Internet has really changed the nature of competition in this market, and has also had an effect on angle and curve of the lines in Figure 2-1.

In this chapter, we focus on the practical level of service elements (Heskett, Sasser et al., 1997; Normann, 2000) that ICT-enabled marketing channels deliver, and performance of such a channel[3] performs with regard to these service elements (e.g. a customer will compare the various channels in terms of the ease and quality of 'information provisioning' or 'decision support' they offer).

A useful service element set that we created from several sources is: problem recognition/ clarification or need identification (Dobler and Burt, 1996); information provisioning (Hill and Alexander, 2000); decision support (Harink, 1997; Hill and Alexander, 2000); generating trust (Grönroos, 1994; Hennig-Thurau and Hansen, 2000; Steinfield, Wit et al., 2000); negotiation (Stern, El-Ansary et al., 1996); ordering; payment; order processing; pick, pack, ship & delivery (Bowersox and Bixby-Cooper, 1992; Stern, El-Ansary et al., 1996; Lynn, 2000); transferring the possession of goods (Stern, El-Ansary et al., 1996); inventory management; communication to co-ordinate activities (Bowersox and Bixby-Cooper, 1992; Stern, El-Ansary et al., 1996; Lynn, 2000); notifications (which decrease the cost of lost opportunities) (Heijden and Valiente, 2002); calamity support & distress relief (Anton, 1996, 2000); after sales usage support; and relationship building (Hennig-Thurau and Hansen, 2000; Lynn, 2000).

Table 2-1. Service elements of marketing channels	
1. Need identification	10. Transferring possession of goods
2. Information provisioning	11. Inventory management
3. Decision support	12. Coordinate tasks, track & trace, order status
4. Generating trust	13. Notification (to support needs)
5. Negotiation	14. Calamity support and distress relief
6. Ordering	15. After sales usage advise & support
7. Payment	16. Relationship building/maintenance
8. Order processing	
9. Pick, pack, ship & delivery	

The service elements can be placed in a logical context and structure. One of the most frequently used perspectives is that of the sales cycle (Gebauer and Scharl, 1999). There are several ways of phasing the cycle; we have chosen the simple form of pre-sales, sales and after-sales (Bhattacherjee, 2001), see also Table 2-2. As can be expected, there are limitations to this model. Not all service elements are limited to one phase: elements like providing information (3), generating trust (4), communication for the co-ordination of tasks (12) and relationship maintenance (16) can typically be placed across the entire cycle.

---

[3] Since we are looking for the potential value of channels, we focus on the properties of well-organized channels. For example, a poorly managed call center with long waiting times will frustrate users and perform poorly on ease of access, even though well-managed call centers can be a very convenient way of contacting an organization quickly.

Table 2-2. Extended service element list, incl. customer & supplier requirements

Service element	Customer requirements	Supplier requirements
	Pre-Sales	
1. Need identification	Obtain non-intrusive suggestions to help me understand & fulfill potential needs	Generate sales opportunities / leads
2. Information provisioning	Find out whether a product or service is good for me	Educate customers about my products/services
3. Decision support	Obtain help in making choices that are right for me	Reduce uncertainty to stimulate purchase
4. Generating trust	Check if supplier, offer and processes are trustworthy	Instill trust/confidence, reduce uncertainty
5. Negotiation	Fit offer to my wishes	Close deal at limited process & price costs
	Sales	
6. Ordering	Purchase & initiate transfer of possession	Receive order
7. Payment	Pay conveniently and safely	Receive payment
8. Order processing	Process order 'invisibly'	Efficient order handling
9. Pick, pack, ship & delivery	Obtain goods timely and easily	Minimized logistics costs & adequate delivery
10. Transferring possession of goods	Obtain possession	Transfer possession
11. Inventory management[4]	Availability of products/service within easy reach	Minimized inventory/capacity costs of acceptable coverage
	After-Sales	
12. Co-ordinate tasks, track & trace, order status	Co-ordinate tasks and ensure effectiveness	Efficient and effective co-ordination of tasks
13. Notification (support needs)	Support my tasks or help seize opportunities	Stimulate usage, service or repeat sales at minimal costs
14. Calamity support and distress relief	Have urgent/important problems resolved & damage repaired	Recover image and relationship
15. After sales usage advise and support	Resolve questions and problems, increase usage benefits	Support usage and satisfaction to increase future revenues
16. Relationship building/maintenance	Increase ordering efficiency and effectiveness by better knowing suppliers	Increase repeat sales by binding & knowing customers

Our sales cycle service element model can be used as a reference to describe auxiliary services offered by channels. A priori, it is hardly possible to say which elements are more important than others, and to what degree. In contrast, according to design literature, one of the key challenges in any service design is

---

[4] Inventory management relates to tangible products, and is a concept that originates from the strong logistics focus in traditional channel management. An equivalent for services could be 'capacity management', which aims at maximizing service capacity within easy reach of customers at minimal costs.

determining which service elements are the most important for the targeted offer and customer group. Moreover, to offer a competitive customer experience, alignment and coherence of the multi-channel service system are important, including adequate adoption of new technologies. According to service literature, partial design and local sub-optimization are among the highest risks in designing and implementing a total service offer and service system (Ramaswamy, 1996; Johnston, 1999; Grönroos, 2000; Normann, 2000). For this reason, rigorous design support methodology can be expected to aid the design process and quality (Clausing, 1994; Ramaswamy, 1996). These are some of the reasons from theory for further developing design support methods in our research.

The initial phases of the design process involve requirements engineering, the goal of which is to transform user requirements that exist in a natural language into formal specifications to serve as a basis for design and implementation. However, most techniques are geared mainly towards building data and process models and focus on technical aspects of the information systems to be developed (Herzwurm, Schockert et al., 2002). Moreover, their philosophy is *elicitation* of user requirements, while it is more a process of service *creation* that has to be supported (Ramaswamy, 1996; Gordijn, 2002). In chapter 3 we look for methods that are close to natural language and that can support service definition processes for e-services in a multi-channel context.

Next, we want to introduce a distinction between customer and supplier requirements per service element. The reason for this is that channels help make a market. They aim at matching supply and demand, which is why some of their service elements will be aimed more at serving customer needs and others more at 'supplier needs'. The requirements of both parties have to be considered if one is to optimize a channel mix, as is illustrated in Table 2-2, which lists customer and supplier requirements for each of the service elements. Note that the requirements are phrased in an active form (using verbs), indicating what customers and suppliers want to achieve in using/ providing a service element, in everyday language. This follows an accepted practice of user centered design (Ramaswamy, 1996).

The split between customer and supplier requirements shows that certain service elements incorporate a potential conflict of interests: e.g. need identification (1), generating trust (4), negotiation (5), pick, pack, ship and delivery (9), inventory management (11), and all after sales services. The interests of supply and demand are not always the same. From this it can be anticipated that a certain channel implementation can be geared more to serving suppliers than customers, or vice versa. Hence, when one is designing multi-channel servicing, both the customer's and the supplier's needs should in some way be taken into consideration.

2.2   Case Study Methodology

The research question of this chapter was what determines the competitive value of a channel within a multi-channel mix, and where opportunities for the Internet can be expected. And although the theory section provides us with an overall framework, we still do not know what the specific (dis)advantages are of different channels in relation to each other. As stated in 1.4 we did not find much literature previous to the start of our research on (differences in) service capabilities of the channels we were interested in. Hence we adopted an explorative case study approach to obtain the comparisons between the channels we were interested in (Internet, call centers, physical stores and personal sales representatives). In the first half of 2000 a team of ten researchers[5] conducted 19 case studies among Dutch firms that had developed an Internet channel with the aim of realizing synergies with existing channels. Several of these cases were revisited early 2002.

We chose the cases to ensure variability across industry/product type and firm size. The specific criteria for case selection included:
- Retail or other physical presence: Selected firms had a physical presence in the Netherlands, either in the form of retail outlets, sales personnel, or field representatives. We did not include dot.com firms because they claim they do not use physical channels.
- E-commerce: Selected firms had initiated an e-commerce activity that appeared to make use of both physical and ICT-enabled channels.
- Product/Industry: Selected firms included those selling physical goods (perishable and non-perishable, durables and small items), and information goods (like books and CD's) and services (like telecom subscriptions and financial services). We also targeted goods and services currently experiencing a greater amount of Internet-based sales, such as financial and travel services, books, music and electronics equipment.
- Size: Wherever possible we selected a large, multiple location firm as well as a small and medium-sized enterprise in each product/industry area. In some cases both of the firms we selected were large, but one of them clearly had a secondary position in the market.

They were selected on the basis of news and trade journal reports highlighting their e-commerce undertakings combining physical channels and ICT-enabled channels. The cases included a large network equipment manufacturer, a consumer and business retail chain of two large telecom operators, a large mobile telecom operator business center chain, a large company selling office products, a large grocery chain, a Health food Internet portal (SME) and a health food store (SME), a large automobile import and dealer organization , a single location automobile dealer (SME), a bicycle wholesaler (SME), a single location bicycle retailer (SME), a large music retailer chain, and a single music store (SME), a large nation-wide financial services/banking provider, a large nation-wide book retailer chain, a smaller upscale book retail chain, and a small multiple location book retailer (SME), an international travel operator and a small travel agency (SME).

With regard to methodology, we collected company information from publicly available sources and reviews of the firms' websites to make sure that their case would contribute to our research goals. We conducted multiple face-to-face interviews with the managers who were most responsible for electronic commerce

---

[5] This research was conducted in the PLACE project, see also Chapter 1.

and channel co-operation, using a structured interview protocol (see place.telin.nl, for a more extensive description of our case study approach, see also (Steinfield, Bouwman et al., 2002)) from March to June 2000. Due to the exploratory nature of the subject the interviews were semi-structured. We used open-ended questions to guide the discussion about the channel development of the various firms. We asked the interviewees how the various channels were positioned in relation to each other, and what the results were of the choices that had been made, and why and how positioning choices were made, what they thought of those choices with hindsight, and what they saw as key future issues. These questions provided us with insights on the relative service value of channels in relation to each other, which is the research question of this chapter. Furthermore, we asked questions on the types of outcome they experienced, internal organizational issues the company faced in introducing Internet-based and other ICT-channels, and the specific applications employed by the firm that involved integration of physical channels and ICT-enabled channels. These questions provided design process and service development information in answer to the next research question, which is addressed in chapter 3. Finally, case descriptions of 8-10 pages were made, on the basis of which cross-case analyses were performed to find similarities and differences, and identify a number of best practices (Yin, 1994). To ensure internal validity, all researchers followed the same interview and case description templates, co-operated on multiple cases in different interviewer combinations, and reviewed and discussed each others' case descriptions. Case descriptions were reviewed by the participants and some of the questions raised in the cross-case comparison were also answered in a second round of (telephone) interviews.

2.3   Case Study Results

Our case results will be discussed in three steps. Firstly, we will list the multi-channel service attributes[6] for each case (Table 2-3), after which we provide a summary for all the cases and relate the results to the service elements presented in Table 2-4.

Table 2-3 lists the results of our multi-channel case study, analyzed according to the service attributes for each channel. We included only those service attribute-channel combinations where a specific channel had particular strengths (+) or limitations (-) in the table. Mere average performance is not depicted. In our approach, the subjectivity of our interviewees is an important given; their

---

[6] One service element can involve multiple customer requirements and multiple service attributes. It is common design practice to distinguish between requirements and attributes, which aids against 'jumping to solutions' and helps to keep design options open. Service or product attributes are the characteristics or properties of products and services. They can be divided in functionalities or functions on the one hand (like 'order entry' or 'provide product information') and non-functional qualifications on the other hand, which usually define quality levels (like 'reliability of a service', 'user friendliness of a Web site', or 'hardness of materials'). In this chapter functional and quality attributes are sometimes still taken together in summarizing our case findings. In the remainder of our research we focus on the functions, since those must be determined first. Quality specification comes later in the service definition process and is outside our scope.

experiences, perceptions and judgments regarding channel (dis)advantages form the basis for our analysis. Consequently, our ratings have been based on the number of positive and negative remarks by interviewees about channels, as well as the degree of positivity/negativity. Note that many more strengths than limitations are mentioned. This is a direct consequence of the opportunity driven focus of channel marketers.

Table 2-3. Examples of the strengths and weaknesses of channels per case				
Type of company	**ICT-enabled and personal communication Channels**			
	Internet	**Call (center)[7]**	**Retail Stores**	**Personal sales**
Network Equipment manufacturer (B2B, tangibles, specialty goods) (see also (Simons, 2001))	(+) easier ordering & product configuration (+) self service trouble shooting (+) downloading invoices (+) view order status, incl. track & trace (+) support product & price comparisons with competitors (+) support dealers with order history mgt information	(+) instant handling of complex questions (+) supporting 'novice' buyers (+) lead qualification (-) less control over support process (-) limited problem overview for customer	[planned as a future channel for SME's, but not yet developed]	(+) advise & educate the customer (+) negotiation
Telecom Operator & Retailer (B2C & B2B, tangibles, commodity and specialty goods)	(+) instant mutations (e.g. subscription type, address) (+) overview product information (+) continuous insight in bill (esp. for cell phone users) (-) delayed product possession	(+) instant handling of complex questions (+) convenience ordering (for less Internet-savvy customers) (+) resolve complaints (+) lead generation (-) less control & overview for customer (-) delayed product possession	(+) allow in store pickup & paying (trust) (+) immediate possession of products (+) experience look and feel of products (+) product returns, repair & replacement (-) limited stock size and portfolio breadth for less popular goods	(+) (complex) deal making and negotiation (+) relationship building (+) product education & advise on use (+) problem clarification (-) too 'heavy an instrument for minor issues

---

[7] Some, but not all, call support is routed via a call center facility.

Company	Internet	Call (center)[8]	Retail Stores	Personal sales
Large Travel company (B2C, intangibles, commodity and specialty goods)	(+) pre-sales info & education (+) comparison of alternatives (esp. for experienced travelers) (+) efficient order processing (+) messaging regarding order status	(+) convenience ordering (for less Internet savvy customers) (+) instant product or status information & decision support	(+) allow in store pickup & paying (trust) (+) more in-depth information & decision support	[not present]
Office Supplies Wholesaler (B2B, tangibles, commodities)	(+) efficient order processing (connection to order management system) (+) easy product comparison (+) instant product availability check	(+) allow for impulse buying (professionals on site, using cell phone) & impulse inquiries (+) resolve complaints	(+) allow product pickup (decreasing use due to lower total cost & high reliability of the alternative: 'home' delivery) (+) experience look and feel of products	(+) relationship building (+) negotiation of window contracts incl. terms of delivery (+) advise on & promotion of online ordering (+) practical help on setting up online account
Large Bank (B2C & SME-B2B, intangibles, commodity and specialty goods)	(+) support routine transactions (+) pre-sales info & education on complex products (+) generate leads (-) handling of complex leads requires other channels	(+) fast response on leads from Internet (+) appointment scheduling for customers and local offices (coordinating tasks) (-) handling of complex leads requires face to face contact	[The distinction between retail and personal sales was blurred for this case: retail personnel would make very personal appointments and make 'home' visits, thus functioning as field sales] (+) face to face contact (generate trust) (+) decision support for high value (financial) products (+) advice on and promotion of use of Internet self help	

Legend: (-) = limitation, and (+) = strength of channel – author's interpretation of interviews

We only highlighted 5 out of the 19 cases. We only selected the most interesting and diverse cases in relation to multi-channel, which means that at least three of the four channels we targeted (Internet, call, retail stores, personal sales) were used extensively. Two of the other 19 cases were very similar to our second case and would add little to this comparison. The other 12 were mostly bi-channel cases, using mainly Internet plus retail stores, or Internet plus personal sales. In some of those cases calling was present (as an option to call a store to order books, CD's, etc), but it was not really used as a separate channel or as part of a deliberate multi-channel service design.

In Table 2-4 we display the summarized service attributes across cases, linked to the service elements. This provides a general overview of what channels can contribute, but the weight and relative advantage of service attributes vary per case and customer (group). For example, the convenience of Internet ordering is relatively high in cases where customers are online all day.

---

[8] Some, but not all, call support is routed via a call center facility.

Table 2-4. Channel service attribute strengths and weaknesses per service element

Service Elements	ICT-enabled & physical channels			
	Internet	**Call**	**Retail Stores**	**Personal Sales**
Pre-Sales Phase				
1. Need identification	(+) lead generation (-) handling of complex leads requires other channels	(+) lead generation (+) fast response on leads from Internet (+) lead qualification (-) handling of complex leads requires face to face contact	(+) in-depth advise	(+) problem clarification (-) too 'strong' instrument for minor issues
2. Information Provisioning	(+) educate customer on product, alternatives etc (+) instant product availability check	(+) instant product or status information	(+) experience look and feel of products (+) in-depth information	(+) educate customer & advise on product use (+) practical help & advise online service
3. Decision support	(+) support product & price comparisons	(+) supporting 'novice' buyers (-) limited control & overview for customer	(+) in-depth advise	(+) support complex or high value purchase
4. Trust			(+) in store product pickup & paying (+) face to face contact	(+) face to face contact
5. Negotiation				(+) negotiation (e.g. window contracts)
Sales Phase				
6. Ordering	(+) easy ordering & product configuration	(+) instant ordering (e.g. when on the move)		(+) support complex or high value purchase
7. Payment			(+) in store paying	
8. Order processing	(+) efficient order processing			
9. Pick, pack, ship & delivery	(-) delayed product possession	(-) delayed product possession	(+) fast possession of products	
10. Transferring possession of goods	(-) delayed product possession	(-) delayed product possession	(+) fast possession of products	
11. Inventory management			(-) limited stock for less popular goods	

After-Sales Phase				
	Internet	**Call**	**Retail Stores**	**Personal Sales**
12. Co-ordinate tasks, track & trace, order status	(+) view order or payment status, incl. track & trace (+) instant mutations (subscription, address) (+) support dealers with order history mgt information	(+) instant inquiries		
13. Notification (to support needs)	(+) messaging regarding order status			
14. Calamity support & distress relief		(+) resolve complaints		
15. After sales usage advise & support	(+) self service trouble shooting (+) downloading invoices	(+) instant handling of complex questions (-) less control & overview for customer	(+) product returns, repair & replacement	
16. Relationship building/maintenance				(+) relationship building (-) too 'heavy' an instrument for minor issues

Legend: (-) = limitation, and (+) = strength of channel – author's interpretation across cases

Also, in our B2B cases the advantages of in-store product pickup were relatively low, because of the costs associated with the time needed for transportation. Online pre-sales information scores best in the cases where the buying decision is information-intensive (e.g. financial services, complex technology products). In our B2C travel cases some interesting social behavior became apparent with regard to pre-sales information: at daytime customers would surf and print information (often at work), then presumably discuss the information over dinner, and order at night: online ordering showed a peak around midnight.

Also, within the cases there are different service attribute valuations: e.g. Network Equipment customers that are online all day consider Internet ordering very convenient, but service technicians that are offline most of the day have an Internet threshold. Or in the Telecom case, with regard to instant product possession advantages, if a consumer needs a new telephone immediately, he or she may prefer to go to the store. However, if he or she still has an old back-up at home and little time to go shopping or there is no store nearby, the delayed product possession associated with online ordering may not be a problem. One of the biggest challenges our facing the firms in our cases was deciding which channel attributes to deploy, based on an assessment of the relative value to their customers, and on overall service coherence.

One of the first things that stand out in Table 2-4 is that there are open spaces, which are different per channel and per service element. This indicates that channels are complementary to a certain extent, and there are clear differences between them. Starting with the Internet, customer benefits are mostly related to pre- and after-sales information and decision support, in combination with 24/7 service access and a sense of control over the (self)-service process. This is especially true for information intensive processes like researching mortgages or technical self-help. Various cases showed that customers have a preference for the 'non-pushy' and 'non-intermediated' nature of self-service. In one case the slower but self-managed searches were explicitly preferred over interpersonal searches (Simons, 2001). The main disadvantage is delayed product possession. For suppliers, low cost self-service opportunities and process efficiencies are the most prominent advantages. Also, there are some low cost options for lead generation and customer education, although the handling of complex leads requires other channels. Interestingly enough, the main issue regarding complexity is customer perception: we have seen highly complex products being bought routinely over the Internet by people who had experience and were familiar with those types of products.

For the call channel the main customer benefits are: instant service access (with even less delay than an Internet log on and search) and advice and support in pre- and after-sales. The human aspect is particularly important in case of calamity support and distress relief, see service element 14 in Table 2-4. Especially when service restoration and empathy are needed, calling is fast, cheap and effective. Customer disadvantages are: limited control over the service process and (again) delayed product possession. Supplier benefits include lead generation and support, and being able to reach and support customers over long distances.

The retail store channel remains important for the physical aspects of buying: experiencing the look and feel of products, having immediate product possession, and also being offered facilities for product returns and repair. Interpersonal aspects like providing more in-depth decision support and creating assurance and trust are important. Supplier benefits are stimulation of (additional) purchasing via uncertainty reduction, and closeness to customers, which creates a certain lock-in. For fast-moving consumer goods (e.g. groceries) the fact that stores carry inventory is a supplier advantage in the fact that the store contains the inventory, and self-service logistics (product pickup by customers) is the most efficient and cheap way of servicing customers . A disadvantage from a supplier's point of view is the fact that a physical cannot match the product range that e-tailers can offer online.

Interestingly enough, our case summary indicates that the personal sales channel is not ideal for the sales and after sales (except relation management) service tasks; the other channels are more effective and/or efficient, and using personal sales only creates delays and/or overkill (except when the relationship is at stake). As far as presales is concerned, however, it is the strongest channel. Customer advantages are: presales support, need identification and personalized advice. Supplier benefits are: lead generation, relation management and penetration (by being able to reach below the surface), and stimulating / facilitating the use of other channels for tasks that can be performed more efficiently there. An important supplier disadvantage is the high cost of personal sales.

2.4 Analysis

Our results have to be interpreted with some caution. First, Table 2-4 is the result of theory generation, rather than theory testing. Hence, there is a question as to the external validity of these results. Moreover, we found that many service strengths and weaknesses depend on customer needs and context. This is confirmed in another study (Bouwman and Wijngaert, 2003). In this chapter, we have focused on the general service attributes of multi-channel servicing, and on a generalized overview of what channels can contribute to service elements in terms of service attributes. Another limitation is that we took a adopted a relatively functional approach to service, which meant that we did not include marketing notions like emotional value of experiences or image effects. Although our multi-channel overviews are in line with our bi-channel case results, there is a difference: firms that use fewer channels are limited in the options they have at their disposal, which will have some impact in the relative advantages of the channels will shift somewhat (e.g. when personal sales is excluded from the channel set, retail may be the best remaining channel for negotiation or personalized advise).

Contrary to recent expectations that space and time would become less important in the information age, our research indicates that some aspects are very important to multi-channel service choices. For example, the Internet, call centers and personal sales are hampered with regard to physical service aspects and are associated with delayed product possession, which is exactly where retail stores score well. However, with regard to product range and efficient inventory management retail stores are more limited. Also, the perceived convenience of a certain channel will depend on the proximity of retail outlets. Thus, when defining a multi-channel service mix, an organization must address these space- and time considerations and how they are valued by customers. This falsifies the simple dotcom era statement that online offers will win the entire market.

From services literature we know that customers generally appreciate control over and involvement in the service process (Johnston, 1999; Grönroos, 2000; Normann, 2000). Comparing service attributes of the Internet and call centers has made some of these differences explicit (Table 2-4). Customers appear to perceive less psychological pressure and more process control with Internet self-service, creating Internet preferences in terms of certain informational and decision tasks.

There are several ways to assess the quality of Internet sites. Some approaches focus on ease of use, completeness of functionality (Verhagen, Vries et al., 2001), visual appeal, or a combination of several aspects (Barnes and Vidgen, 2001; Chen, Gillenson and Sherrell, 2002). However, some of the more successful click and mortar approaches we have seen would not score particularly high on these elements (e.g. our Network Equipment case would only score well on functionality and our Office Supplies case only on ease of use). Our results show that the value of channel functionality depends less on closeness to an objective 'golden standard' and more on the quality with which service design has been performed, translating specific supplier opportunities into supplier benefits and addressing specific customer needs of a segment with suitable service support (Ramaswamy, 1996; Grönroos, Heinonen et al., 2000).

In our theory section, we have separated customer and supplier requirements. The prudence of this choice is confirmed in two ways. Firstly, by our findings that supplier and customer interests may vary and that they may have a different perception with regard to channel solutions. Secondly, with regard to the service development process, we observed difficulties with dealing with tradeoffs and deciding between alternative solutions in several cases.

Based on our findings from the case studies we have tried to transcend the solution-oriented terminology used by channel marketers and offer building blocks for a multi-channel design support method. We have seen that taking a design approach to multi-channel marketing transcends some of the current omissions in Internet-performance and Marketing Channel literature: the terminology and focus are relatively solution-oriented. If we look at, for example, service element 9 - 'pick, pack, ship & delivery' - which has as requirements: 'Obtain goods timely & easily' (for customers) and 'Minimized logistics costs & adequate delivery' (for suppliers), clearly the requirements expose underlying needs more explicitly than the solution-phrased service element.

We do not expect all firms will benefit in the same way from service design methods. With regard to design choices made on multi-channel service concepts, there were significant differences between the various cases. Generally speaking, SME's put very little time and resources into analysis and design. Their focus was on 'just do it'. In several cases, selling books or bicycles, little thought was given to how to make money from Internet initiatives. The fact that customers were asking for Internet links was enough reason. This created loss generating sites. Within the group of large corporations, we found several instances of partial design and sub-optimized or conflicting solutions. One of our telecom cases, for example, chose a consumer-oriented Internet shop as a basis for online service processes, thus hampering the support of business to business processes surrounding order handling. Or a large bank did generate Internet leads, but the local offices were neither ready nor equipped to pick up those leads. The medium-sized companies were generally best at maintaining an adequate multi-channel coherence, and the balance between analysis and quality of implementation. Our case observations indicate that this is mainly due to the fact that the overall service offer across channels could relatively well be managed by one person or team.

The Internet has given customers more control, knowledge and power. This creates a challenge for some of the larger firms we studied to become more customer-oriented than they thus far needed to be, and to address customer requirements explicitly in service design. Paradoxically, it also means that supplier benefits have to be guarded more explicitly, since customers are less easily locked into a sales cycle. When offering services on the Internet for lower prices, or sometimes for free, there is a risk that these low-margin services will not be compensated by consumption of high-margin goods.

2.5   Conclusion

The research question of this chapter was what determines the competitive value of a channel within a multi-channel mix, and where opportunities for the Internet can be expected. We now present the main conclusions, based on explorative case study research across 19 Dutch cases from selected industries, product types, and firm sizes with regard to the adoption and implementation of the Internet in addition to existing channels.

> **Case Exhibit 2-2 Value of channel services shows variations across cases**
>
> The weight and relative advantage of channel services varies per case and customer (group). For example, the convenience of Internet ordering is relatively high in situations where customers are online all day. In our B2B cases the advantages of in-store product pickup were relatively low, because of the costs associated with the time needed for transportation, which contrasts with several of our B2C case findings. Online pre-sales information scores best in the cases where the buying decision is information intensive (e.g. financial services, complex technology products). In our B2C travel cases some interesting social behavior became apparent for pre-sales information: at daytime customers would surf and print information (often at work), then presumably discuss this information over dinner, and order at night: online ordering showed a peak around midnight.

Firstly, channels vary considerably in what they contribute to a channel mix. The strengths and weaknesses of each channel become visible when evaluated per service element across the buying cycle (see also Table 2-4). Secondly, our results show that suppliers and customers have different perceptions as to the strengths of a channel. The points of view of both parties need to be taken into account when one is designing new e-services. Thirdly, an important finding from this chapter is that we should always consider the local context of a case when defining new service opportunities for the Internet. On the one hand, the added value of a new channel should explicitly be evaluated relative to the existing channel mix. For example, people who live close to a supermarket may be less interested in the possibility to order online, whereas people who live further away may appreciate the option. On the other hand, the value for a customer depends on customer context and preferences, which also vary per case, and even among groups within a case (see also Case Exhibit 2-2).

Thus, identifying opportunities for the Internet and Internet service value only to a certain extent is a generic task. There is always a serious design challenge, which involves a case-specific analysis of the 'design problem' and then moves towards developing suitable solutions. This chapter also aids in specifying part of what the design challenge is and what characterizes 'suitable solutions': we have shown that those solutions will have to address both customer needs and supplier needs, as well as add value within the overall channel mix. This design challenge is addressed in the following chapters.

Moreover, in this chapter we developed a conceptualization of the problems at hand, which we summarize here. We used the literature review in this chapter to integrate several approaches. Firstly, we borrowed several key concepts from services marketing (Heskett, Sasser et al., 1997; Grönroos, 2000; Hennig-Thurau and Hansen, 2000; Normann, 2000). The main idea is that the Internet and other channels primairly provide 'auxiliary' or 'peripheral' services, in addition to selling a 'core product or service', which is the good that people purchase and for which they pay. Auxiliary services are generally part of the package: they do influence competitive attractiveness but are not paid for directly. Secondly, we positioned our approach in relation to the well-known SERVQUAL instrument (Parasuraman, Berry et al., 1985; Parasuraman, Berry et al., 1993). SERVQUAL is mostly aimed at measuring appreciation of the 'soft process' (Venetis, 1997) of treating customers with care and integrity, and at the 'tangibles' variable which is part of the 'hard process' (Venetis, 1997) or 'technical quality' (Grönroos, 2000) that expresses some of the functionality and 'serviscape' that are present to serve customers. We explained why this instrument does not tell us which service elements to include and how to include them. Thirdly, we used channel economics and channel marketing literature (Bucklin, 1966; Bucklin, 1972; Stern, El-Ansary et al., 1996; Lynn, 2000) to determine a reference point for evaluating the added value of new Internet channel services. This reference point is competitive advantage in relation to competitors and existing channels. Fourthly, we extracted a set of service elements from literature in relation to sales cycle support for customers (Bowersox and Bixby-Cooper, 1992; Stern, El-Ansary et al., 1996; Gebauer and Scharl, 1999; Bhattacherjee, 2001). This set was useful as a framework for comparing the relative value of channels. Fifthly, we addressed organizational and service coherence challenges for multi-channeling (Ramaswamy, 1996; Heskett, Sasser et al., 1997; Johnston, 1999; Grönroos, 2000; Normann, 2000). This literature shows that it is very difficult, especially for large organizations, to provide a coherent and high quality experience to customers across all channels, and to manage channels in such a way that they offer a coherent contribution towards the same goals. Finally, we briefly connected to design approaches (Clausing, 1994; Ramaswamy, 1996; Gordijn, 2002; Herzwurm, Schockert et al., 2002). One of the important lessons from this literature is that it makes sense to separate the design phases of problem analyses and solution synthesis. This helps avoid a premature 'jumping to solutions'. This is also the background for moving beyond 'service elements' to user requirements on the one hand versus service attributes and potential solutions on the other.

Based on this conceptualization, we have come to define the introduction of an e-service as a design problem. Within that design philosophy, we view service definition as a marketing and competitive positioning challenge. Thus, we attempt to integrate services marketing and service design. Furthermore, based on our case study observations we have shown that channel (dis)advantages depend on customer segments, type of supplier, type of market, customer context and the overall set of channels being used. Hence, we illustrated that the biggest challenge is choosing what to offer to whom: defining a proposition that adds value to customers as well as suppliers and in relation to existing channels. This can be seen as a true design problem with a marketing perspective.

3   Design Processes, Challenges & Methods

Specialization: knowing more and more about less and less, until one finally knows everything about nothing
(Anonymous)

Case Exhibit 3-1 Illustration of e-service development difficulties: Large Travel Company

This case illustrates the early stages of an e-business approach. Moreover, uncertainty prevails in this case. Large Travel Company mainly consists of several retail chains selling leisure travel to consumers. A significant part of activities in the branch offices is related either to order entry or to handing out brochures to customers to help them in their orientation. Both activities provide limited added value over what customers themselves can do on the Internet. Nevertheless, apart from one small Internet-based brand within the organization, most retail chains choose to largely ignore the Internet. There are some people in the central organization who think differently, but this is not enough to create changes. People are moving in different directions, IT and the multiple marketing groups are not in sync. Customer orientation or market orientation are limited, especially in the existing branch offices. There is resistance to change, there is no shared image of which Internet services to implement or how, and as a result there is no focused service development effort.

In this chapter we answer our second research question[9]:

2. How can design methods aid the process of designing new e-services that have to function in a multi-channel context, and what are the requirements for such a design method?

To answer this question, we take several steps. In theory section 3.1 we start by discussing methods from services marketing and from more general design theory. Then we extract requirements for multi-channel design support methods from theory. In chapter 2 we concluded that it is a serious challenge for organizations to define value adding propositions for a) customers as well as for b) the supplier and c) in relation to existing channels. The outcome quality requirements we identify in this chapter are directly related to this challenge: 1) customer-oriented design, 2) channel coherence, 3) channel synergy, defined as

---

[9] This chapter is based on an article that was published in the International Journal of Electronic Business, Volume 3, Issue 1, 2005. An earlier version of this article was presented to the 5th World Congress on the Management of Electronic Commerce, Hamilton, Ontario, Canada, January 14-16, 2004.

re-using assets across channels (Power, 2000), and 4) competitive positioning. In addition, we identify four design *process* quality requirements: 5) speed, later operationalized as 'progress, 6) focused design process, 7) stakeholder communication, and 8) communication of service concept coherence during implementation. In section 3.2 we use case studies to research how the development of e-services next to existing channels has taken place in practice, and to substantiate our requirements further. After that we discuss our findings and evaluate the design methods we reviewed in relation to our design requirements in 3.3. In section 3.4 we draw our conclusions. Case Exhibit 3-1 illustrates some of the difficulties in developing new e-services in addition to an existing channel arrangement.

3.1   Theoretical background

Although service design has been identified as 'perhaps the most crucial factor for quality', it is one of the least studied topics in services marketing (Gummesson, 1993). In this section we present a brief outline of the methods and requirements found in service and product design literature. This outline of design methods provides a basis for our further e-service definition method development. In section 3.3 we evaluate the design methods in relation to our outcome and process requirements. And in chapter 4 we develop our method on the basis of that evaluation and the design methods we introduce below.

Methods in services literature

There are several studies that discuss existing service design literature (Tax and Stuart, 1997; Johnston, 1999; Goldstein, Johnston, Duffy and Rao, 2002; Menor, Tatikonda and Sampson, 2002). They all agree that service design methodology has not yet been fully developed. In practice, services are often poorly designed. It has been said that if we designed cars the way we do services, they would have three wheels and four axes.

The most methodical and design-oriented (as well as the most frequently cited) approach to services marketing is called service blueprinting (Shostack, 1984). It is a graphical technique for flow-charting service processes, both above and below the 'line of visibility', indicating the processes that are visible to customers and those that are not. Service blueprinting helps identify potential failure points and makes it easier to consider various countermeasures. Generally, defining regular service operations is less demanding than defining processes to deal with exceptions, failures and countermeasures (Hull, Jackson and Dick, 2002). Service blueprinting helps identify the impact of new services on the existing organization. For a new service to be successful, it is important to be aware of this kind of impact (Tax and Stuart, 1997).

At a broader, organizational level, a method called 'service system planning' can be used. The overall service system is made up of: (1) the customer, including needs and expectations, (2) the service concept, (3) the service delivery system, (4) the image (which influences short term service delivery expectations of servers and customers) and (5) corporate culture and values (which determines the long-term service orientation of an organization) (Normann, 2000). At a more detailed

46

level, the service delivery system contains the following components: (a) the roles of people (i.e. service providers and customers), (b) technology, (c) physical facilities, (d) equipment (used in the service process by servers or customers), and (e) service delivery processes (Heskett, Sasser et al., 1997). The design and evaluation of new services that are added to a company's portfolio can be aided by looking at the various service system components, and asking how they are (or should be) affected. Although this method is less well developed in terms of customer orientation and design process management, it does provide a useful checklist of the main components involved.

At the level of strategic service marketing, a method can be used whereby the service concept is defined explicitly. The service concept integrates the *how* and the *what* of a new service (Goldstein, Johnston et al., 2002). On the 'what' side the service concept connects a company's strategy to what is important to customers. As such, the service concept can serve as a high level reality check: is this the right service for us to be designing? On the 'how' side it connects the detailed service functionalities of designers to the overall service experience that is intended. For example, a "day out at Disney's magic kingdom is more likely to be defined by its designers and its visitors as a magical experience than six rides and a burger in a clean park" (Clark, Johnston and Shulver, 2000). Detailed service definition tasks run the risk of ignoring the fact that customers tend to see a service in its entirety. An explicit service concept definition can be a countermeasure against that risk. Although by itself a service concept is not enough to manage the design process, it does provide a direction as far as the outcome of the design is concerned.

Product design methods, fundamental engineering and total design
Since product design has a longer tradition than service design, more design methods are available in that area. Services have only been recognized since the 1970's and 1980's (Johnston, 1999) as being fundamentally different from products in a number of ways (most notably the fact that the production and consumption of services generally take place at the same time, which is rarely the case with goods).

Methods to support product design have similar aims: enabling decision-making, widening the space for potential solutions and facilitating teamwork. Some designers tend to view these methods as 'straightjackets', although in most cases they do in fact serve as 'lifejackets' that help a design team stay afloat. The design approach displayed in Table 3-1 consists of several tasks. Those tasks can be categorized according to their aim and stage in the design process (Cross, 1994). Reading from the top of Table 3-1, we move from the overall problem to sub-problems (phases 1, 2, 3), from sub-problems to sub-solutions (phase 4), and from sub-solutions to overall solution (phases 5, 6, 7). We refer to this approach as 'fundamental engineering'. For a full description of the approach used, see Cross (1994).

Although fundamental engineering is certainly not without its merits, one of its basic drawbacks is that as a design process it is separated from marketing and operations. But customer-, marketing- and operational considerations should be an integral part of service definition decisions to avoid suboptimal designs

(Clausing, 1994). Hence, a more integrated, comprehensive approach is preferred. We will refer to this as 'total design', in contrast to what can be described as 'partial design'.

Table 3-1 Overview of the 'fundamental engineering' design approach (Cross, 1994)

Design phase	Description of phases
Clarifying objectives	Clarify design objectives and sub-objectives, and the relationships between them.
Establishing functions	Establish the functions required and the system boundary of a new design.
Setting requirements	Make an accurate specification of the performance required of a design solution.
Determining characteristics	Set targets for the engineering characteristics of a product, such that they satisfy customer requirements.
Generating alternatives	Generate the complete range of alternative design solutions, and hence widen the search for new solutions.
Evaluating alternatives	Compare the utility values of alternative design proposals, on the basis of performance against differentially weighted objectives.
Improving details	Increase or maintain the value of an offer to its purchaser, whilst reducing its cost to its producer.

According to literature, partial design and local sub-optimization are among the highest risks in designing and implementing a total service offer and service system (Ramaswamy, 1996; Johnston, 1999; Grönroos, 2000; Normann, 2000). 'Partial design' or 'fundamental engineering' is described as the type of design that is taught in technical schools, and whose main emphasis is on setting up formal requirements and technical solutions, but which has a limited focus (Clausing, 1994; Ramaswamy, 1996). Both in terms of focus and of the language being used, this approach is relatively technical, which makes it difficult to discuss the proposed services with (potential) customers. In fact, it often even makes communication between 'marketing' and 'technology' quite difficult. Moreover, the focus is on product/service design, which means that issues concerning elements like ease of production or operations with regard to the market support side of a service are often overlooked.

Similar problems are found in the discipline of 'Systems Engineering,' which is generally used to develop software systems or 'software products' like aircraft dashboards. A drawback of systems engineering is its focus on the more technical aspects of information systems. The models are created for specialists, and customers can only understand parts of them (Herzwurm, Schockert, Dowie and Breidung, 2002). One of the more customer-oriented methods is the development of 'use cases' describing end-user scenarios and functions. These can be used to model system behaviors in UML, which are used for software specification. Use cases are similar to service blueprinting, although they are often used at a more detailed level. We discussed service blueprinting in our service methods review.

Compared to these more technical approaches to design, total design takes a broader view: all the design elements are ultimately based on customer needs and requirements, while other factors, such as ease of production, are also taken into account. Total design comes from the Total Quality movement, where total quality is described in terms of customer orientation, overall design and production robustness, and the prevention of failures rather than the need to repair them. Total design promotes the use of rigorous methods to improve design process and quality. Examples of total design are Concurrent Engineering and Quality Function Deployment (QFD) (Mizuno and Akao, 1994; Clausing, 1994). One the one hand, concurrent engineering is a way of thinking - "concurrent engineering simply means that the systems engineering process should be done with all of the phases [..] of the system life cycle in mind" (Prasad, 1996; Buede, 2000) - while on the other hand it is a design approach, a way of (team) working, making sure that all the relevant disciplines (including expertise on how to manage, deliver and correct services in 'real life') are included in the design team.

QFD is a method that consists of a set of systematic tools designed to support a structured way of concurrent engineering. It was developed to deal with some of the major problems in traditional design processes, such as
- Disregard for customer needs,
- Disregard for competing offers (and channels),
- Concentration on each specification (and channel) in isolation,
- Little input from design and operational people in terms of product planning,
- Lack of structure,
- Loss of information across design phases,
- Low level of commitment to previous decisions (Clausing, 1994).

QFD is a rich tradition that has produced many detailed design tools, for an overview see Chan and Wu (2002). We focus on the QFD core, as the details lie outside the scope of this chapter. In QFD, as in fundamental engineering, the requirements and attributes of the design are separated explicitly. What QFD adds is a list of customer priorities, in everyday language. These priorities are a point of reference throughout the design process. All choices (and trade-offs) are viewed in terms of their impact on customer priorities and potential solutions are assessed on the basis of how well they score in this respect compared to the solutions of competitors. To work out the design in greater detail a sequence of matrices ensures that the right priorities are kept in mind all the way down to the levels of sub-processes, data entities and subsystems (Mizuno & Akao, 1994; Cohen, 1995; Ramaswamy, 1996; Herzwurm, Schockert et al., 2002), see also Appendix A. QFD assesses design attributes (e.g. 'compare products on selected features' or 'no delivery charges') in terms of their adherence to customer priorities. In short, QFD is customer-oriented (Ramaswamy, 1996), it addresses competitor performance, and it is a rigorous, highly visual and integrating method, using fixed tools and formats (Clausing, 1994).

Requirements on multi-channel design support
In this section we make the requirements explicit for evaluating design methods (Simons and Bouwman, 2006b). As design support often aims at generating certain *outcomes*, we begin by taking four requirements from theory, focused on generating better outcomes. Many services are unsuccessful because their design lacks *customer orientation* (Ramaswamy, 1996). With regard to multi-channel servicing it is important to avoid channel conflicts that are to the result of poor alignment of Internet services within the overall service design: what is

49

needed is *channel coherence* (Simons, Bouwman et al., 2002; Simons and Bouwman, 2004). Channels have to work together if they are to maximize overall customer value in such a way that the strengths of each channel are used and the various channels complement each other. A seamless and consistent customer experience across the channels will evoke customer trust, which will reinforce the relationship. The aim is, therefore, to develop an e-service that properly accounts for the service value provided to customers by physical channels in relation to the new e-service. Moreover, channel coherence is important to make it easier to show what their role in the overall mix is (to customers and internally to employees). This enhances quality of service delivery on a day-to-day basis (Johnston, 1999; Grönroos, 2000; Normann, 2000; Wootten, 2003). From a cost savings perspective it is also crucial to strive for *channel synergy effects* between the various channels. We adhere to the narrow definition of channel synergy (Power, 2000) that focuses on reusing assets to minimize costs. During the dot.com hype too many initiatives started under a new brand, targeting new customer segments, using separate order entry systems, separate logistics, separate information systems etc, with all the additional costs involved. More integrated click and mortar approaches were often more successful, partly due to lower cost levels (Steinfield, Wit et al., 2000; Simons, 2001; Simons, Bouwman et al., 2002; Steinfield, Bouwman et al., 2002; Simons and Bouwman, 2003a). Finally, the roles and functions of channels need to match the marketing strategy of the service provider. If a company's employees fail to see the coherence and synergy of the channels being used, it is highly unlikely that the customers will (Grönroos, 2000). Also, services have to be designed to support a value added market position as well as offer value for customers (Steinfield, Bouwman et al., 2002; Simons and Bouwman, 2006a). In short, a new e-service has to contribute to *competitive positioning*. Hence the following four outcome-focused requirements on design support methods have been shown to be relevant for e-services in a multi-channel context: customer orientation (Ramaswamy, 1996), channel coherence and channel synergy (Normann, 2000; Power, 2000; Wootten, 2003), as well as contribution to the supplier's competitive positioning (Steinfield, Bouwman et al., 2002).

Next, we consider the quality of the design *process*. Time to market and *design process speed* are important issues anywhere, but in service design people tend to give themselves especially little time (Gordijn, 2002). A *focused design process* is necessary to avoid the problems of traditional design (such as low commitment to previous decisions, loss of information, etc) we addressed in the previous section (Clausing, 1994). *Communication across stakeholder perspectives* is important in early design phases to avoid biased solutions. In particular, customer contact channels run the risk of being driven by only one of the functions (sales, customer service, marketing, or IT). This compromises the quality of solutions (Power, 2000; Rigby, Reichheld and Schefter, 2002). Finally, bridging design and implementation is always a challenge (Cohen, 1995; Ramaswamy, 1996). With customer contact services this is important, since from the design stage onwards the new services have to show their value via the day-to-day operations of 'blue collar' employees. Those employees have to be able to promote and explain those services to customers at the 'moments of truth' when they contact a firm. Hence, they have to understand the main design choices behind those services: it is important to *support concept coherence and communication during implementation.* Hence, with regard to the quality of the design process itself, the following four requirements are important: design process speed (time to market), a focused design process, multi-disciplinary team communication, and support for concept coherence and communication during the implementation phase (Clausing, 1994; Cohen, 1995; Ramaswamy, 1996; Gordijn, 2002).

In the next section we use case studies to specify and substantiate the design process requirements for outcome and process related design requirements. And in section 3.3 we use the substantiated requirements to evaluate the design methods described above.

3.2   Case Study

In section 2.2 we introduced our explorative case study among 19 Dutch cases from selected industries, product types, and firm sizes regarding their adoption and implementation of the Internet along existing channels. For the purpose of this chapter, all case reports were subjected to a *qualitative* content analysis in which all the statements were selected and categorized regarding the design requirements that were related either to the outcomes of the design process or of the process itself. The initial general categories - the four process requirements that address outcomes and the four design process requirements - were used as sensitizing concepts and relevant quotes were attributed to these categories.

In this section we begin by describing the design process. In the next section we apply the requirements we identified to the design methods from our literature review in an effort to find out the extent to which those methods meet the requirements.

Multi-channel Design Processes

As a first step, we asked the interviewees how the various channels were positioned in relation to each other, what choices had been made and why, what the results of those choices were and what the interviewees thought of those choices backing retrospect. As far as the introduction of electronic services is concerned, we found that many companies shared a similar experience. Common themes were: creativity, speed, chaos, having to deal with uncertainties, significant organizational changes, and innovation-centered or solution-driven design and implementation. The various companies experienced these common themes at different levels of intensity, and the strategies they used to deal with these challenges showed a marked variation. One thing the companies did have in common, however, was that none of them used a well-structured design process with transparent or explicit decision-making rules aimed at dealing with major dilemmas. This was especially true in the early stages of the development, when important choices had to be made. In those cases where a design approach was used, it was only further down the design process, at the point where the e-service concept was handed over to the technicians to prepare implementation.

Table 3-2 Design process comparison across multi-channel cases

Type of company	Multi-channel service design process
1. Network equipment manufacturer (B2B, tangibles, specialty goods)	Very mature e-business approach; Early adopter of Internet channel; high speed and flexibility in design and implementation; much learning by doing; First e-services were based on solving a number of explicit 'headaches' of customers and channel partners Centralized decision making; close cooperation between marketing and IT; Close monitoring of customer satisfaction, channel satisfaction, marketing effectiveness and financial performance.
2. Telecom Operator & Retailer (B2C & B2B, tangibles, commodity and specialty goods)	At the time of case study still in early stages of e-business approach; Started as many early decentralized Internet initiatives; Centralization caused problems: lost functionality, customer dissatisfaction, less cross-channel support and technology-driven implementations; No coherent channel implementation or management of customer contact points.
3. Large Travel Company (B2C, intangibles, commodity and specialty goods)	At the time of case study still in early stages of e-business approach; Poor internal support regarding Internet-based activities; much internal resistance; no focused development effort Existing physical channels not customer-oriented or market-centered in their approach; Low speed in design, decision-making and implementation.
4. Office Supplies Wholesaler (B2B, tangibles, commodities)	At the time of case study still in early stage of e-business approach; Good match between customer focus and market strategy (many shared goals and principles); Close cooperation between all channels and relevant disciplines (marketing, IT, fulfillment), One central Internet initiative team with sufficient authority to make and implement decisions, as well as an adequate span of control due to limited size of the firm.
5. Large Bank (B2C & SME-B2B, intangibles, commodity and specialty goods)	Mature e-business approach; High-level marketing strategy aimed at channel coherence, although service design and implementation based on company preferences rather than customer requirements; Poor coherence and cooperation between Internet and branch offices. Despite positive intentions many leads were dropped.
6. Mobile Operator & Retailer (B2C & B2B, tangibles, commodity and specialty goods)	At the time of case study still in early stages of e-business approach; No clear idea of what the role of each channel is; Limited insights into customer needs; No competitive benchmark or Internet functionality benchmark generated.

In Table 3-2 we compare cases where at least 3 out of 4 channels (Internet, call, retail stores, personal sales) are used extensively. The Network Equipment case presents the most structured and coherent example of multi-channel service design we found. The first e-services were based on solving a number of explicit 'headaches' of customers and channel partners. All e-services ideas were evaluated in terms of customer and supplier advantages. The role and added value of the Internet were made explicit, as well the roles of other channels. Problems and uncertainties, e.g. regarding customer preferences and user adoption of new services, were addressed in an experimental fashion, with short feedback loops to learn and improve. Regarding management of the new service development process, the clear and centralized decision-making structure helped, as well as extensive communication of what was being done and why.

Only our Office Supplies case comes close to the structure and coherence of the Network Equipment case. One central e-commerce team managed the coherence of the channel mix and added value of new e-services. This team was involved in other commercial activities and had sufficient authority to implement decisions. Moreover, the e-commerce team combined all the relevant functions (IT, marketing, sales, customer service). E-service goals were formulated in detail and communicated to both customers and internal organization. Compared to the first case, the scope and complexity of the e-services and parties involved were much smaller. Therefore, it is easy for one team to control and execute all Internet initiatives. In our first case, control and communication were centralized, but exploration, definition, implementation and management of new e-services took place elsewhere, namely within specific business units.

The Telecom Operator & Retailer and Large Travel Company struggled above all with a lack of centralization: different people were moving into different directions, there was no shared goal, and sub-optimization occurred across business/product lines and across functions: IT and marketing.

The Large Bank and Mobile Operator and Retailer suffered mainly from insufficient clarity on e-service concepts: what should be offered online, and why, and how should e-service channels co-operate with other channels? As a consequence, in all four of these cases, there was no orchestration of customer value or channel cooperation, which hampered service implementation and competitive strength.

In the remaining thirteen cases, only one channel was used alongside the Internet. This simplified the marketing and design problem. Still, the new service development processes we observed were still seriously challenging those firms. The size of the company proved important: SME's generally put very little resources and time into analysis and design. The only exception we saw was the call center and Internet-based Small Travel Shop, which did put a great deal of effort in developing inter-channel synergy. Generally speaking, SME's with a physical presence tend to spend less time and effort in managing a professional Internet presence. Large firms were often unclear with regard to what the customer needs or indeed the competitive forces were and they showed poor internal coordination, sub-optimization and conflicts of interests. Generally speaking, medium-sized companies proved best at maintaining multi-channel coherence, and balancing analysis and quality of implementation, mostly because it was relatively easy for one person or team to manage the overall service across the various channels. In overall terms there was some degree of 'technology push', with marketers and technicians alike wanting to implement many features and functions at the same time. When asked why, they replied: 'Because we can and because it is nice to do.'

Design Support Requirements
As a second step, we used the requirements we found to assess the various types of design support. Based on our case analysis, we believe that in particular large organizations would benefit from multi-channel design support methods. The teams responsible for developing and implementing e-services are faced with a number of tasks: identifying and evaluating technical possibilities, assessing customer needs and service adoption, assessing the (strategic) value for the supplier of e-services, assessing conflicts or synergies between the Internet and other channels, and planning/guiding implementation. Design support focuses on the design process and helps teams focus on specific outcomes. Four design process requirements contribute directly to better design *outcomes*, while four other requirements directly support the design *process itself*. Below, we specify and substantiate the requirements with regard to multi-channel service design, based on what we found in the case studies.

Process requirements aimed at design *outcomes* quality:
1   *Customer-Oriented Design*: Many services provided by channels are 'additional' (i.e. they are not included in the core 'product') and non-billable. Customer value has to be incorporated into the channel service design. Clear decisions have to be made regarding which service features to include in the total service offer.
2   *Channel Coherence*: The customer value provided by other channels in addition to e-services has to be taken into account. E-services must be developed that complement other existing channels.
3   *Channel Synergy*: Channel synergy is the re-use of assets across channels. Re-use of assets must be stimulated while maintaining customer focus in designing the value proposition and service processes.
4   *Competitive Positioning*: The starting point is an existing organization with a specific channel structure, a specific type of customers and a specific set of organizational capabilities. Hence, a multi-channel solution that suits one firm in a specific market does not necessarily suit its competitors. Choices must be made as to what to retain, as a competitive asset or differentiator, and what to discard.

Design *process* quality requirements:
5   *Speed*: e-service concept exploration may only take a limited period of time, typically a few weeks. A fast but thorough and systematic exploration is needed.
6   *Focused Design Process*: Although the number of available service options and channel combinations tends to explode, it is vital that the design process remain focused.
7   *Communication between Stakeholder Perspectives in Development Team:* A wide range of stakeholders is involved, ranging from CEO's to channel

specialists, representatives from sales and service and from IT & processes. A wide range of interests and perspectives must be supported while maintaining focus.

8   *Communication of Concept Coherence during Implementation:* During implementation, the balance between customer and supplier needs, and between one channel and the others, is easily lost. The coherence of the e-service design must be maintained and communicated during implementation and operation.

These more refined design requirements describe the generic needs of design teams responsible for the development of e-services in a multi-channel context. For example, cases that were both customer-oriented in their design and had a good fit with the marketing strategy of the supply side were rare. However, in most of the cases we studied at least one of those two requirements was not seriously considered. Many interviewees also indicated that incorporating channel coherence and channel synergy into the service design was challenging. Time to market was important in all cases, hampering formal design procedures. In all cases the issues covered multiple disciplines like marketing, technology and operations. Thus, it is important to enable communication and focus across those disciplines in the design process. Finally, in a number of cases, the implementation was left largely to the 'technicians', which meant that no explicit connection was made between customer needs and the proposed solutions. As a result there were service failures (which damaged customer relations), channel conflicts (which damaged channel effectiveness) and unsuccessful or loss-generating service implementations (which led to a negative return on investment).

3.3   Discussion

In this section we evaluate design support methods based on the requirements identified in the case studies, using the Pugh method (Clausing, 1994; Cohen, 1995; Ramaswamy, 1996), a method aimed at assessing various solutions (in our case: design support methods) on the basis of a set of requirements (in our case: design requirements). One of the more promising solutions is used as a reference point or 'datum,' with the other solutions being scored as better (+), worse (-) or same (S) in relation to that reference point on all requirements, shown in Table 3-3.

Following accepted engineering practice, the Pugh scores in the table are the result of our assessment of the solutions in relation to the requirements. The aim is to achieve inter-subjective consensus on the relative performance of the various solutions. This is also why difference-scores to datum are used (instead of comparing 'absolute' scores of solutions). This creates a more focused discussion, rather than having to wonder whether a score should be a 3 or really a 4 (Clausing, 1994; Cohen, 1995; Ramaswamy, 1996). The scores in Table 3-3 are a summary of our subjective evaluation, to make our judgments explicit. Below, we briefly explain our scores, before going on to discuss their implications. Another limitation in relation to Table 3-3 has to do with the external validity of the requirements. Our list of requirements is the result of our literature study plus exploratory research across 19 cases. Our results show that these requirements cover relevant multi-channel design issues that are generic across our case firms, but this has not been validated by other cases.

The first method we discuss is blueprinting. Blueprinting scores best on 'speed' since it is a fast and intuitive method for outlining what happens or should happen in the relevant customer processes. Compared to QFD it also scores better with regard to 'channel coherence'. The reason for this is that combining customer processes and the company resources involved in supporting these processes helps identify related tasks in the non-Internet-based channels. Blueprinting scores worse than QFD on 'customer orientation' and 'design process focus', as do the other methods. None of the other methods share the attention QFD pays to customer's needs and requirements. Finally, blueprinting also scores worse than QFD on 'concept coherence during implementation'. This is mainly due to the fact that blueprinting fails to provide the detailed tools needed to guide implementation.

Table 3-3 Evaluation (via Pugh method) of design methods against QFD method

Design requirements	QFD (datum)	Blueprinting & Line of Visibility	Service system planning	Service concept definition	Fundamental Engineering
Customer-oriented design		-	-	-	-
Channel coherence		+	+	S	S
Channel synergy		S	+	S	+
Competitive positioning		S	S	+	-
Design process speed		+	S	S	S
Focused design process		-	-	-	-
Stakeholder communication		S	-	S	-
Concept Coherence and Communication during Implementation		-	S	S	-

Legend: + = better than QFD; S = same as QFD; - = worse than QFD

The second method we scored is service system planning. Service system planning scores better than QFD on 'channel coherence' & '-synergy', because it explicitly addresses the total service system. QFD focuses more on a single service concept and service element detail; as such it pays less attention to organizational structures. Service system planning scores worse than QFD on 'stakeholder communication', because the design process is not managed as it is in QFD. Moreover, it does not facilitate the communication between marketers and technicians. Finally, there are negative scores on 'customer orientation' and 'design process focus', as explained in the previous section.

The third method we evaluated is service concept definition. This method scores better than QFD on 'competitive positioning' because the marketing strategy and strategic discussions are explicit reference points. QFD is aimed more at service tactics and operations. The service concept definition method scores worse on 'customer orientation' and 'design process focus' requirements. Service concept definition scores the same as QFD on 'stakeholder communication' and 'concept coherence during implementation', although that score is based on other strengths. QFD is more rigorous and provides more detailed guidance for all the disciplines and implementation steps involved. The service concept definition method provides a summary of the service concept which is easier to communicate: i.e. what are we accomplishing and why?

The fourth method we scored is fundamental engineering. This approach has a positive score on 'channel synergy'. Synergies are likely to be found because improvement of details (see Table 3-1) focuses on opportunities to save costs and enhance value. It has a negative score on 'customer orientation' and 'design process focus', as well as on 'competitive positioning', 'stakeholder communication' and 'concept coherence during implementation', due to the fact that both the language and the method being used are of a technical nature. The fundamental engineering methods provide very few explicit links between marketing, engineering, production and servicing.

Based on our evaluation we can conclude that QFD has some advantages as well as drawbacks. Some of the advantages are related to the degree of customer orientation it provides by connecting all design decisions to customer priorities (first requirement). As far as the other three design outcome requirements - enabling channel coherence, channel synergies and competitive positioning of the supply side - are concerned, QFD has several limitations. It is more suitable for optimizing a specific service implementation than for incorporating elements such as strategic interests, the behavior of other channels, or developing a market strategy. With regard to the first process requirement, i.e. supporting design speed, QFD's limitations become most apparent. The steadfast and methodical nature of QFD would appear to make it less than suitable for something as chaotic as conceptualizing a new e-service. In its traditional form, QFD is perhaps too rigorous and detailed an approach, aimed as it is at robust design (Clausing, 1994; Cohen, 1995; Ramaswamy, 1996). This contrasts strongly with the hasty and 'sloppy' approach to service design that we observed in our cases. As far as the last three process requirements - focused design process, and communication across disciplines during and after the design phase - are concerned, QFD shows a relatively high performance in comparison to other methods. This can be attributed to the decision-making and communication structures it provides.

To summarize, we conclude that a leaner version of QFD, which maintains a number of its tools (to capture customer priorities, translate them to solutions and focus communication throughout design and implementation), is most suitable to support early multi-channel design choices. Having said that, we feel that it is important to modify QFD to incorporate multi-channel considerations and to make it suitable for the fast-paced field of e-service design.

3.4   Conclusion

In this chapter we answered the research question - a) how design support methods can aid the process of designing new e-services that have to function in a multi-channel context, and b) what requirements are for such design support. In summary the answer to our research question is that a) a lean, fast version of

QFD, with multi-channel additions, can be expected to aid the process of designing new e-services, and that b) requirements for design support are: 1) customer-oriented design, 2) channel coherence, 3) channel synergy, 4) competitive positioning, 5) speed, 6) focused design process, 7) stakeholder communication, and 8) communication of service concept coherence during implementation.

Moreover, our case studies revealed that a majority of companies fail to use a structured approach to e-service development, which affected the quality of decision-making. This state of affairs resulted in service failures (thus damaging customer relations), channel conflicts (which damaged channel effectiveness) and unsuccessful service implementations (with negative return on investment). Based on our case findings, we choose to focus on the initial definition phases in the remainder of our research: the 'initiation' phase in terms of Alter (1999). We expect that it is useful to aid decision making in the initiation phase by developing a systematic e-service definition method that guards firms against one-sided e-service initiatives, but instead incorporates customer orientation, multi-channel considerations and supplier interests. Our focus on the initiation phase also implies that we will not evaluate the method we develop with regard to the eighth requirement - communication of concept coherence during implementation - since this has to do with the implementation phase.

We reviewed several methods from the services marketing and product design traditions. These methods were evaluated on the basis of the requirements mentioned earlier. None of the methods scored well on all of the requirements. Moreover, we found that design methods for services are scarce. In conclusion, the existing service design methods prove useful for visualizing service interactions, speed, providing service system descriptions, and decision-making with regard to service concepts and strategy. They are less useful in terms of rigor, customer-orientation and the evaluation of service alternatives, areas where product design methods may prove useful, having proved their value elsewhere.

In all, our conclusion is that a leaner version of QFD, which maintains a number of its tools (to capture customer priorities, translate them to solutions and focus communication throughout design and implementation), should be the basis for our e-service definition method which is suitable for a multi-channel context. It is important to modify QFD to incorporate multi-channel considerations and to make it suitable for the fast-paced field of e-service design. An important constraint following from our case research is that the method has to be modified to become much faster, to connect to the impatient practitioners of service development in general and of e-service development in particular. This is addressed in the next chapter.

Part II: Test Phase

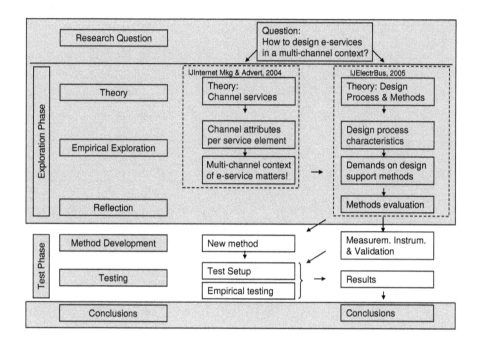

4 Development of multi-channel design method

Whether you believe you can do a thing or not, you are right.
(Henry Ford)

Case Exhibit 4-1 Pretest of first version of MuCh-QFD illustrates desire for speed and focus

We pre-tested a first version (without GroupSystems™ support) of our MuCh-QFD session with a large financial firm (banking and insurance services) in February 2004. Two parallel groups, of 4 and 3 participants respectively, used a preliminary version of the MuCh-QFD approach for e-service design in a multi-channel context. The groups were of a multi-disciplinary nature, with participants from operations, marketing, finance and ICT. The intake phase was carried out in a preceding session with representatives of the firm, who were asked to formulate three e-service possibilities. From this, an e-service was chosen that provides online support for mortgages based on mutual funds. This case was considered suitable because of the high support needs of customers and the importance of intermediaries in servicing customers (multi-channel aspects).

The main finding of the pretest was that the session tasks were neither concise nor precise enough. There was too much room for different interpretations. This led to significant differences between the two groups in terms of design process and in terms of outcomes. For example, group A described the customer process in much more detail than group B, and used less detail for the functionality of each of the process steps. This finding was confirmed in the individual and group evaluations. Participants wanted more clarity with regard to the tasks they were asked to carry out and they complained that many issues, although they had been discussed, were left undecided. Thus we decided to give much more rigor and focus to our second version of MuCh-QFD.

The goal of this chapter is to answer our third research question:
3. How can we develop an e-service definition method that meets our design support requirements?

To answer this question we take several steps. In the previous chapter we concluded that QFD was likely to provide a useful basis for our e-service definition method. Hence, in section 4.1 we first describe several key elements of the Quality Function Deployment (QFD) design approach, several of which we will use in our new method. Then we explain our method in section 4.2, and the relationship between QFD and our new 'Multi-Channel QFD' (MuCh QFD) method. We present our conclusion in section 4.3. Case Exhibit 4-1 illustrates the need for focus and speed in supporting e-service definition processes, a need explicitly stated by business participants.

4.1   Background information on QFD

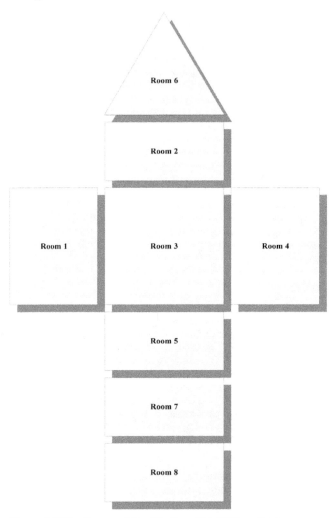

Figure 4-1 The House of Quality from the QFD method

Quality Function Deployment (QFD) was developed in Japan in an effort to encourage engineers to consider quality early in the design process (Mizuno and Akao, 1994). 'Quality Function Deployment' is a direct translation of the characters *Hin Shitsu, Ki No, Ten Kai*, the phrase meaning something like "the strategic arrangement (deployment) throughout all aspects of a product (functions) of appropriate characteristics (qualities) according to customer demands". Some of the key words to capture the spirit of QFD are: attaining high quality, preventing design flaws, addressing customer needs, and stimulating communication across the organization and across the product life cycle.

QFD started in the Kobe shipyards and was further developed by the Japanese automotive industry, after which it spread across the globe. QFD started as a product development approach and in the past ten years it has been used increasingly in service design (Mazur, 1993, 1995; Ramaswamy, 1996; Herzwurm, Schockert et al., 2002). The QFD method uses the 'House of Quality', see Figure 4-1, which is a matrix-based system for organizing the design process. In the QFD philosophy a multi-disciplinary team should fill the 'House of Quality to ensure incorporation of marketing, technology and operations issues'.

Below we discuss the rooms of the 'House of Quality' and in Figure 4-3 we illustrate how the rooms are used.
- *Room 1 – The "Whats" Room*: Customer needs in the verbatim form of the 'voice of the customer' and a rating of their importance are collected in room 1. In Figure 4-3 the two columns on the left show an example of customer needs plus importance ratings, as developed with one of our participating cases.
- *Room 2 – The "Hows" Room*: This room contains the 'voice of the design team' and lists the service functions (or product attributes in case of product design) that the design must have to satisfy the customer needs (Ramaswamy, 1996; Herzwurm, Schockert et al., 2002). In Figure 4-3 the top row shows an example of service functions. Some service functions carry the same name as one of the customer needs. This happens more often in service design than in product design, and is not a problem (Ramaswamy, 1996). It reflects the nature of what services do: providing processes and functions that fulfill customer needs.
- *Room 3 – The Relationship Matrix*: Room 3 is the relationship matrix, where the associations between service functions and customers needs are represented. The goal is to identify which service functions contribute the most to the main customer needs. Via group discussion, the team assigns a value to the "what/how" relations. The values could in principle have any range the team chooses. Standard practice in QFD is to focus on 'strong' relationships, which are assigned a value of 9, and 'medium' relationships, which are assigned a value of 3[10]. The 'strong' relationships are considered most important for design choices; hence the QFD convention of using 9's versus 3's increases their relative weight. A team has the option of identifying 'weak' relationships, which are assigned a value of 1, but the impact of the '1'-values is limited. In Appendix A we further explain the QFD rules of thumb for filling the relationship matrix. In Figure 4-3 the large 'middle' matrix filled with 9's, 3's or with empty cells is an example of a relationship matrix.

---

[10] Note: This convention differs from the terminology used in statistics, where a correlation between 0.9 and 0.7 is called 'strong', between 0.7 and 0.4 'medium' and between 0.4 and 0.2 'weak'.

- *Room 4 – Competitive Benchmarking*: Benchmarking with competitors is carried out in Room 4 to obtain customers' views about the extent to which their needs are met by competing products or services. This is illustrated in Figure 4-4. Normally room 4 connects to the right side of room 3 (which is indeed the case in the MS Excel based templates we use), but we have chosen to leave out the relationship matrix from the figure for ease of representation.
- *Room 5 – Technical Benchmarking*: This is the technical counterpart of Room 4. The design team evaluates the quality of the performance of the competitors' designs, based on performance variables specified for each service function. The practice of how to specify performance variables (and norms, see Room 8) falls beyond the scope of our research. Interested readers are referred to Cohen (1995), Ramaswamy (1996), Chan and Wu (2002) for further information.
- *Room 6 – The Roof*: Here correlations between service functions are identified. A negative correlation indicates potential design conflicts or trade-offs. For example, 'simple product overview' may provide a trade-off with 'compare products on multiple details'. It is up to the design team how it wants to deal with this trade-off. A positive correlation, on the other hand, indicates opportunities where service functions may help reinforce each other's value if they are designed together consciously.
- *Room 7 – Importance Rating*: Here the importance of each service function is calculated from the data on the importance of customer needs Room 1 and the associations in Room 3. This information helps the team identify the service functions that best meet customer needs. In Figure 4-3 this is shown as the bottom row where all functions are given a weighted score, depending on customer priorities. Although the scores should not be taken too literally, they do provide insight into the importance of service functions based on customer needs.
- *Room 8 – Performance Norms*: In this room the expectations for the new service (or product) are quantified by defining performance norms for each of the service functions. Room 5 - technical benchmarking in relation to competitors - provides guidance with regard to the values that will characterize a superior product or service.

In the next section we focus on the way that HoQ rooms are used in our MuCh-QFD method. We refer the reader to other works (Clausing, 1994; Cohen, 1995; Ramaswamy, 1996; Chan and Wu, 2002; Herzwurm, Schockert et al., 2002), but we will limit ourselves to the 'basics' described above as a basis for the remainder of this thesis.

4.2   Design of 'Multi-Channel QFD' (MuCh-QFD) method

In this section we describe how we arrived at our 'Multi-Channel QFD' (MuCh-QFD) method. In chapters 1 and 3 we explained that we focus on the support for the 'initiation' phase (Alter, 1999) of the design process. We start this section by further specifying the scope, design approach and service definition tasks we address in our method, after which we explain the format of our method in relation to the design requirements. This format is a four hour formalized e-service definition session with multiple stakeholders, which is supported by a formalized intake. Hence as next step we provide a brief outline of the topics addressed during the intake and in the session. Finally, we describe the links between QFD and the agenda of the MuCh-QFD session.

The scope of MuCh-QFD is limited to the initial design phases. In Alter's (1999) terminology, we focus on the 'initiation' phase, which precedes the development and implementation phases. The initiation phase addresses the purposes and goals of the e-service idea, scope and feasibility, and functional specifications. Another approach (Kar, 2004) distinguishes the following phases: analysis, preparation, synthesis, implementation and testing. In terms of this approach we focus on the analysis phase. Within that phase we address service process design (comparable to service blueprinting), customer requirements analysis, existing services analysis (in our case explicitly comparing online and offline services) and functional specification. We thus generate a high-level service description. If we compare our scope to the design phases identified by Cross (1994) as mentioned in chapter 3, MuCh-QFD mainly addresses the first three phases: clarifying objectives, establishing functions and setting requirements (in our case: 'customer requirements'). The fourth phase, determining characteristics, is partly addressed by MuCh-QFD in that it generates functional specifications and provides an overview of the constraints and online solutions.

The next question is how to support speed, focus and communication between stakeholders, and perform the design tasks in a way that supports our outcome quality requirements (from customer orientation and multi-channel considerations to competitive positioning). This involves questions regarding format as well as questions concerning design task content. We start by addressing the format question.

There are two e-service definition process formats that are very common in firms (which we know from our consulting experience and which were also encountered in our explorative case study across 19 cases described in chapters 2 and 3). The two formats are: a) an idea owner (possibly in cooperation with a sponsor at management level) performs the initiation phase tasks, or b) a project team is formed which carries out the e-service initiation tasks, using team meetings interspersed with 'homework' by participants. More often than not, these participants also carry out their 'regular' tasks, and the initiation phase regularly spans several weeks or months. In terms of speed and focus, performance is generally low under these conditions. The 'idea owner' format (a) often limits the active involvement of the main stakeholders. We therefore suggest a third format: a formal e-service definition session, preceded by an intake using a formal intake protocol. Either the idea owner (a) or project team (b) can initiate this session, as long as they involve the relevant stakeholders in the e-service definition process.

The following design process considerations have played an important role in our choice for a session format. Firstly, it is fundamental to the approaches of QFD as well as concurrent engineering to involve multiple stakeholders in the design process as early as possible (including for example marketing managers and sales representatives next to people responsible for technology or operational processes). As discussed in section 3.1, this helps avoid the problems of 'partial design' and sub-optimal service development. Few stakeholders perform design tasks regularly, which is why a four hour session that facilitates a collective analysis and high level functional specification fits their work practices and preferences better than taking part in a formal design project. Secondly, business practitioners do not want to invest much time into initial service definition phases, which is especially apparent when there is a limited commitment for the e-service idea. Hence, we wanted a format that required a limited one-time investment. Thirdly, group sessions run the risk of evoking discussions that do not result in a great deal of progress in terms of concrete e-service specifications, design choices or other outputs. To create specific outputs it helps to have a formal structure, a strict agenda and commitment. In our pre-tests we found that using an electronic decision support system like GroupSystems™ help provide structure during the sessions, create focus on outputs and make progress. Given these three considerations, we decided to use a formalized four hour session supported by GroupSystems™, because it a) facilitates joint e-service development between multiple stakeholders, b) requires a limited time investment and c) generates high level functional specifications based on a formal analysis of the main stakeholder interests (customer-, channel- and supplier interests).

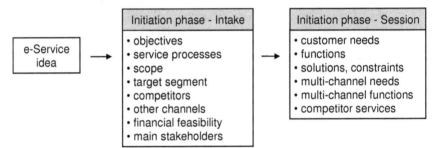

Figure 4-2 Design tasks during intake and service definition session of MuCh-QFD

An important threat to progress and focus is having an e-service definition that is too broad, or having design session goals that are too broad, which is why we developed an intake protocol (see Appendix B for details) to narrow down the scope of the design problem to be tackled during a session. A second purpose of the intake was to carry out some of the e-service definition tasks of the initiation phase. Figure 4-2 summarizes the tasks during the intake and the subsequent e-service definition session. We describe the intake and describe the tasks performed during MuCh-QFD sessions, after we have explained the relationships between QFD and MuCh-QFD.

The intake protocol is conducted with one or two idea owners. It starts by defining the auxiliary e-service idea, the core product that it is meant to enhance and the customer target segment. Then the main service process steps that a customer experiences are defined (following service blueprinting basics, but without explicit 'line of visibility' exercises; question 3 of the protocol). The next step is to determine the relevance of the e-service in relation to market demand and where the focus of the session should lie. This is done in questions 4 and 5 of the protocol. Those questions determine in which phase of the buying cycle the e-services have the greatest customer and competitive value, which is why the design session will focus on that phase. The next step of the intake (questions 6 to 10) discusses multi-channel issues and competitor (e-)services. This step helps us analyze the multi-channel and competitive contexts of the e-service. Questions 11 and 12 create a high-level business case (costs and benefits), as a first check on financial feasibility. Finally, it has to be determined who should be session participants. In other words, who are the main stakeholders that should be involved in the early stage of the service definition process? To ensure customer orientation, multi-channel coherence and synergy, and competitive positioning, there must be stakeholders that represent customer interests - generally this will be someone from sales or marketing (research), the marketing goals of the supplier - generally this will be someone from marketing or management, operations- and IT (im)possibilities - IT and/or process experts, and channel interests - representatives from the main channels that are used in addition to the Internet - generally someone from personal sales, contact centers or physical stores.

When we look at the QFD House of Quality rooms presented in section 4.1, we see that several rooms link directly to our design requirements. The guidelines we used to develop our MuCh-QFD session from QFD were that we wanted to a) use the direct links with HoQ rooms that were available, while b) maintaining a concise session, and c) proceeding as far as possible in the session without requiring the participants to conduct a priori research (thus lowering entry barriers). Following these guidelines, Rooms 1 to 3 contribute to customer-oriented design, which is our first design requirement, and Room 4 contributes to competitive positioning, which is our fourth design requirement. Rooms 5 and 8 relate to technical quality and are excluded, since they require a priori desk research (guideline c) and/or more time than is available during a session (guideline b). Moreover, the link between Rooms 5 and 8 and our design requirements is not very direct (guideline a). Room 6 is excluded because it requires extensive discussion (guideline b) and since an important part of its value is related to the combination with rooms 5 and 8: in what ways do correlations between functions affect the technical performance in relation to competitors (Cohen, 1995; Ramaswamy, 1996), which means that direct links to our design requirements are limited. Because Room 7 is automatically calculated on the basis of rooms 1 to 3, we include it as service function priority feedback to participants. To summarize, we felt that Rooms 1 and 4 would be most valuable to us, and Room 7 is a logical addition.

Since there are no explicit places in QFD to include multi-channel considerations, we developed multi-channel extensions to QFD as described below. In Table 4-1 the activities for the MuCh-QFD session are listed, together with their connection to QFD House of Quality rooms. In most session activities we used a group decision support system (GDSS) of GroupSystems™ to generate and cluster input, and rank clusters based on the votes of all the participants. For details regarding intake protocol and implementation of the sessions in the group decision support system of GroupSystems™, see appendices C and D. The MuCh-QFD sessions are held with multiple stakeholders from marketing, distribution channel organizations, operations and IT, and sometimes from product management, customer service or sales, depending on the e-service at hand.

In the introduction the intake results are summarized by the moderator in the form of a presentation with room for questions, discussions, etc. The presentation addresses the goal of the e-service definition session, the procedures of the day, a short explanation of the design method that will be used, an outline of the e-service idea, why it would be valuable to customers, what the core product offer and the target segment are to which the e-service is connected, comparisons with competitive initiatives, and expected financial benefits and costs. Part of the explanation of the e-service idea is to make explicit which part of the sales cycle is chosen for the e-service definition session. This choice enables speed and focus in the session.

Table 4-1 Relation MuCh-QFD session agenda and QFD 'House Of Quality' (HoQ)

MuCh-QFD session	Connection to QFD HoQ or extensions	Activities
Introduction		
- Short review of the results of the intake		Presentation
Part I: Customer needs and Internet functions		
- Identify, cluster and prioritize customer needs	Room 1, HoQ	1) GDSS brainstorm, 2) Cluster via group discussion, 3) Vote
- Identify and cluster Internet functions	Room 2, HoQ	1) GDSS brainstorm, 2) Cluster via group discussion
Part II: Define e-service matrix		
- Evaluate functions with regard to needs and create an e-service matrix	Rooms 3 & 7, HoQ	Breakout exercise
- Define service slogan that summarizes proposition	Service Concept Definition	Breakout exercise
- Discuss solutions and constraints for functions	Explore solutions & constraints	Breakout exercise
Part III: Tasks of other channels		
- Check the desired support from other channels	Room 1 & 2 for multi-channel service	1) GDSS brainstorm, 2) Vote
- Check win-win options between e-service and other channels	Room 2 (& 6) for multi-channel service	1) GDSS brainstorm, 2) Vote
- Extend matrix with needs and functions related to other channels	Rooms 3 & 7 for multi-channel service	Breakout exercise
Part IV: Competitive position		
- Discuss strong and weak points of the new e-service in relation to competitors	Prepare room 4 & explore constraints	Breakout exercise
- Score existing-, competitor- and new e-service on customer needs and discuss results	Room 4, HoQ	1) GDSS vote, 2) Group discussion

In part I of the agenda - needs and functions - the participants generate a list of the most important customer needs and the Internet functions that would to a certain extent fulfill those needs (the extent to which they do so is made explicit in step II). This starts as a typical brainstorming task. Both the customer needs and the Internet functions are then categorized, after which the customer needs are prioritized by means of a group vote, with the help of the group support system. The average overall vote is used as a weight factor. Participants are advised explicitly to conduct customer research after the session to update and validate the scores. However, for the sake of expedience the preliminary are used.

Part II of the agenda - define e-service - takes the participants away from the GDSS terminals and consists of three elements. The moderator writes the group outputs on centrally positioned flip over sheets to guide and centralize discussions and decision–making process. The team begins by evaluating the extent to which the functions contribute to the customer needs by filling in an 'e-service matrix' that is derived directly from the QFD-method and illustrated in Figure 4-3. The left column contains the customer needs, and the top row the main functions. The extent to which the functions fulfill the needs that have been identified can be written in the cells. Following standard QFD practice, a 9 is used to indicate a strong correlation and a 3 to indicate a weaker but significant correlation (Cohen, 1995; Ramaswamy, 1996; Herzwurm, Schockert et al., 2002). See appendix A for the more detailed description of the procedures. The team then summarizes the e-service proposition for customers in the form of a service slogan to create a shared image of the coherence between assumed customer value and marketing intention. This activity is partly inspired by the Service Concept Definition method (chapter 3). As a third element of agenda part II (define e-service), the main constraints and potential online service solutions are identified for the different functions. This exercise is not exhaustive, which means that not all functions are necessarily addressed. Instead, the team is asked to express what they see as the main issues and potential service solutions that should be addressed later on in the design process. In most sessions several of these service solutions are already expressed while filling in the e-service matrix and the moderator will have inserted them in the appropriate functionality columns as they arose.

Customer needs	Weights	FAQ	Mini-calculation with incentives	Process overview	Calculation function and comparison	Call-back function	Links to and information about providers	Product information	Recommendation complementary products	Cancellation service	Personal interview	Price incentive	Active migration customers
1. Price information and price benefit	9		9		9								
2. Comparing premiums and conditions	8,5				9				3				
3. Quality product	7,5	3		3				9					
4. Quality provider	7,25					3	9			3			
5. Tailor-made advice	7,25						3		9		9		
6. Product information	7	3						9					
7. Insight into process	6,75			9						3			
8. Insight into legal process	4,25	3		9				3					
(9. User-friendliness)	8	3				9				3			
10. Offer me channel that I am used to	7,75					3					9	9	9
11. Need for other products	6					3			9		9		
Importance:		80	81	122	179	135	78	156	119	66	189	70	70

Figure 4-3 Example of MuCh-QFD e-service matrix (rooms 1 to 3 & 7), incl. multi-channel functions in three columns on the right; Importance = $\sum$ weights x scores

Part III of the agenda - other channel tasks - also consists of three elements. First, the team assesses the desired support from other channels, like retail shops, call centers, etc. This is done via a brainstorm in which channel support desires are generated, followed by a group vote, using the group support system, in which the most important support desires are determined. Secondly, the team assesses a win-win potential between the e-service channel and other marketing channels. This again involves a brainstorm session followed by a group vote. The win-win opportunities can either support channel coherence (enhance marketing effectiveness by cooperation between channels) or channel synergy (cost advantages by reusing resources, data, etc across channels). The third element again takes the participants away from the computers, in an exercise designed to expand the e-service matrix with multi-channel needs and functions. Based on the top 3 support needs from other channels and the top 3 win-win opportunities, one or more customer needs and service functions are added to the matrix. This is illustrated in Figure 4-3 as customer needs 10 and 11 in the left column and in the three additional service function columns to the right of the matrix. As a result, the service matrix represents an overall image of the service proposition, with e-service functions as well as other channel functions.

Customer needs	Weights	New	Present	Direct writers	Banks
1. Price information and benefits	9	8	3	7	6
2. Comparing premiums and conditions	8,5	9	5	4	4
3. Quality Product	7,5	7	6	7	7
4. Quality provider	7,25	7	7	7	7
5. Tailor-made advice	7,25	6	7	6	7
6. Product information	7	8	7	8	8
7. Insight into process	6,75	6	6	6	7
8. Insight in legal process	4,25	6	4	5	6
(9. User friendliness)					
10. Offer channel that I am used to	7,75	6	7	5	6
11. Need for other products	6	7	7	8	8
Importance:		497	423	441	461

Figure 4-4 Example of competitive position assessment (rooms 1 & 4); Importance = $\sum$ weights x scores

Part IV - competitive position - is the final phase of the session. The team starts by discussing the strong and weak points of the new e-service in relation to competitors. The main points are written down on a central flip-over sheet and presented to the team members for their agreement. After that, another electronic group vote is carried out to evaluate how the old (e)-service, the new e-service and the competition's e-services score in relation to customer needs. This is plotted as an extension to the service matrix (see also Figure 4-4). Although it adopts the format of 'room 4' of the QFD house of quality and usually extends to the right of Figure 4-3, we have split the figures here for the sake of representation. This step creates overall scores for the different competing e-services, which are weighted for the relative importance of the customer needs. Also, it becomes clear what the relative scores of the various e-services are. At this point we make it very explicit that the scores are not objective and that their reliability is limited. However, they do present a first picture of the various competing offers. This picture is discussed by the team to see if there are any surprises, or if there is anything that should be taken into account when proceeding with this e-service, etc.

By the end of the session, the participants have created a high-level functional e-service definition, including cross channel connections and explicit links to customer priorities and competitive positioning. These results are meant as a basis for the design phase that follows the initiation phase: the development phase (Alter, 1999). Below, we conclude this chapter by reviewing the contribution of MuCh-QFD to the design process.

4.3   Conclusion

The research question we addressed in this chapter was how to develop an e-service definition method that meets the seven design support requirements we identified eaelier (customer orientation to stakeholder communication). In this chapter we placed the e-service definition problem in the initiation phase, which precedes the development and implementation phases (Alter, 1999). The initiation phase addresses the goals of a service, its scope and feasibility, and translated the results of the phase in functional specifications.

The contribution of MuCh-QFD in terms of speed, focused design process and stakeholder communication is that we developed a format (intake plus a four hour e-service definition session with the main stakeholders, aided by GroupSystems™) that has advantages for each of the requirements. It allows for fast decision-making in the initiation phase, with active involvement of the main stakeholders, and uses methods that create focused exploration and decision-making processes. Moreover, in terms of design process quality, the MuCh-QFD approach explicitly confronts stakeholder interests in the MuCh-QFD matrices. This helps justify and communicate e-service definition decisions and stimulates places the e-service definition in its competitive, operational, technical and multi-channel context. We expect that our format will contribute to an active involvement of the various multiple stakeholders because it is more easily integrated in the current work practices of managers, marketers, sales and channel representatives, compared to traditional design projects.

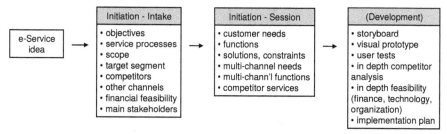

Figure 4-5 Initiation (MuCh-QFD) versus development phase (post-MuCh-QFD)

The contribution of MuCh-QFD in terms of the quality of the design outcome (customer orientation, channel coherence, channel synergy, competitor positioning) is illustrated in Figure 4-5. The intake addresses not only the objectives, scope, customer target segment, market potential and financial feasibility (via a high level business case) of the e-service, it also addresses the main concerns of customers during buying decisions, the roles of other channels, the main competitors and the main stakeholders for the e-service. This helps prepare the more in depth decisions during the session, which is our second step in the initiation phase. In the session the main customer needs are defined and used as an explicit starting point for e-service definition: functions and their importance are explicitly linked to customer priorities. Practical remarks regarding how functions respond to customer priorities (in terms of online solutions like specific decision support tools, and in terms of constraints like 'product comparison should require less than 5 clicks') are also collected. Next, the multi-channel support needs of customers are collected and translated into consequences for multi-channel service functions. Finally, competitive positioning is discussed in relation to customer needs: how does the e-service score in relation to services offered by competitors? As explained in the previous section, the consequences of customer, channel and competitor analyses are linked together in the MuCh-QFD matrices. Making these connections helps increase e-service definition quality.

The next design phase is the development phase, where detailed requirements analyses are conducted. And, according to Alter (1999) the development phase also includes what one might call 'technical development': "hardware, software and other resources are acquired and configured and documentation is made regarding the operation of the information system within the larger work system". Figure 4-5 focuses on the requirements analyses part of the development phase. The results of the initiation phase are use to work out the requirements analyses in greater detail by storyboarding the e-service, making a visual prototype and conducting user tests with that prototype. The visual prototype can also be used for an in-depth analysis of the differences with competitors, and the storyboard can be used to carry out feasibility checks regarding technology, operational processes and finance. Finally, an implementation plan can be made, on the basis of which management can decide whether to proceed to Alter's 'technical development' tasks and to implementation.

5 Development of control group method

If you want to learn how something works, try changing it
(Evolutionary biology)

Case Exhibit 5-1 Second pre-test as illustration of testing MuCh-QFD against control groups

Our second pretest was performed in April 2004 with four groups of master's students with a design background. The pre-test followed a 2x2 setup, testing not only the effect of the design method, but also the effect of using different moderators. The sessions adhered to a strict process, were facilitated with the aid of GroupSystems™ and focused on narrowly specified outputs. Each group was given the same assignment. They were to act as a design team of a mobile phone manufacturer, and define a new e-service. To determine the impact of different experimental conditions, we developed questionnaire constructs for our design requirements (customer orientation to stakeholder communication).

As far as moderator influences are concerned, we concluded that they were minimal. With regard to the impact of the design method, we found that the MuCh-QFD averages were higher than for the control groups on all seven requirements. However, based on Mann Whitney tests, only three differences were significant: customer orientation ($p<0.005$), competitive position ($p=0.01$) and communication ($p<0.05$). These levels of significance are striking, given the small number of observations ($n=16$).

Our case observations have led us to believe that a) MuCh-QFD performed better on the following three requirements: customer centricity, competitive position and stakeholder communication, that b) measurements on channel coherence and speed could become significant when the number of participants would be higher, but that c) using the constructs for channel synergy and focus as differentiators was of limited use. Regarding synergy, several items were not discriminatory; e.g. currently most click & mortar e-services are likely to 're-use existing brand value'. Also, several items needed to be rephrased to increase the focus on multi-channeling and/or on design process observations. As far as focus was concerned, discriminatory measurement proved to be hard. The scores were exceptionally high under all of the experimental conditions. For both methods and both moderators the averages were 7.0 to 7.4 out of 8 items.

In this chapter we answer our fourth research question:

4. Which e-service definition method can we use as a control group condition?

The answer to this question breaks down into several steps. First (5.1) we ask ourselves what type of format the control group design process should follow (for example the 'regular' process a participating firm usually follows, or also a session format)? After we conclude that a session is preferable, we use section 5.2 to consider what type of e-service definition method such a session should use (for example an alternative design method, or no method at all?) and we derive the e-service definition method of the control session, called 'Fundamental Engineering.' Then we describe how the 'Fundamental Engineering' control group session works in detail (5.3), which modifications were made to the MuCh-QFD and FE sessions based on pre-tests (5.4) and which differences we expect to find between MuCh-QFD and FE sessions (5.5). We present the conclusions of this chapter in section 5.6.

5.1 What control group design process to use as reference point?

In this section we answer the question which type of design process we should use as the control group condition. In other words: what are we going to use as a reference point to compare the method we developed in chapter 4 against? There are several possibilities: 1) 'the regular' service definition process taking place in firms, 2) a formal in-company QFD design process, or 3) a similar format design session, based on a different design approach. To help us decide which type of design process to use for control groups we compare the (dis)advantages of all three possibilities. Their (dis)advantages vary in the areas of methodology, practicality and reference point. In Table 5-1 these (dis)advantages are summarized, and they are further explained in the text below.

In terms of methodology (and to a lesser extent practicality) the first two options are not favorable. Using them would mean that, in addition to our MuCh-QFD session, we would have to evaluate a (longer lasting) design process within the same participating firms. Hence, MuCh-QFD would likely influence the control groups (or the other way around if MuCh-QFD is carried out last), which would mean that the experimental and control conditions would influence one another, which eliminates possibilities for conclusions with regard to causality (Hagenaars and Segers, 1980). The third option, using a control group session based on a different design approach, provides a serious methodological advantage: the experimental and control conditions do not influence each other so long as we make sure that participants do not contact each other before both sessions have been completed. Important practical advantages of the third option are that with session formats a relatively large number of case experiments can be conducted within a given time frame and the design process conditions can be controlled relatively well in comparison to 'in situ' processes.

Table 5-1 Evaluation of design process options for control groups

	Methodology	Practicality	Reference point
'Regular' process in firms	(- -) experimental and control conditions are likely to influence one an other (since we apply both conditions to every case[11])	(-) hard to determine start and end of the service definition process (-) fewer cases can be included with the same time investments	(+) 'real world' (-) control condition very different among cases
In-company QFD process	(- -) see above	(-) see above	(-) does not have its reference point in 'real world' service definition processes[12]
Session based on different design approach	(+) easy to control for case-dependent factors by splitting case teams across MuCh-QFD and control condition (+) no interaction between experimental and control groups (-) which design approach to use?	(+) a relatively large number of cases can be used within a given time frame (+) design process conditions can be controlled	(-) does not have its reference point in 'real world' service definition processes (+) control group tasks are well-specified, hence comparisons of design tasks can be relatively precise

Legend: (- -) = severe drawback; (-) = drawback; (+) = advantage

Another important issue is the 'reference point' that is created via the control condition. In the following chapters the (dis)advantages of MuCh-QFD are determined in terms of the differences in results between MuCh-QFD and the control group condition. The advantage of the first option, using 'the regular' service definition processes of case firms as control group condition, is that 'real world' situations are used as a reference point. Options 2 and 3 introduce significantly more structured and methodical approaches for control groups than the regular e-service definition processes we described in chapter 3, which make them somewhat biased as reference points. Hence we are conducting a variation

---

[11] There are important methodological constraints to keep in mind when reading Table 5-1. We will only be able to use a limited number of case firms (many fewer than 20 or 30) for our tests. This means that our experimental design is potentially sensitive to case-dependent singularities: two 'strange' cases are enough to create significant disturbance of our results. As a countermeasure we want to 'split up' our participating case firms, or at least the design teams that participate in our research. This means that each case is subjected to the experimental (MuCh-QFD) as well as the control condition. This limits the impact of case-dependent disturbances, but requires specific protection against influences between the two conditions. If one of the conditions is applied first and is then allowed to influence the effects of the second condition, it becomes virtually impossible to draw conclusions about the causal effects of MuCh-QFD (Hagenaars and Segers, 1980).
[12] The service definition processes we encountered in chapter 3 were on average more chaotic.

on 'the classical experiment'. The variation is that our control condition does contain some design method elements instead of none at all. On the other hand, option 3 makes it possible to control and compare design tasks across sessions with relative precision, which creates the advantage of a well specified reference point.

When we weighed all the pros and cons mentioned above, we decided that option 2 has the largest number of disadvantages, which meant we had to choose between options 1 and 3. Because of the serious practical and methodological disadvantages, we then decided to discard option 1 as well. This meant that option 3 was the only remaining option. Thus, the control group condition we chose next to the MuCh-QFD sessions is a session based on a different design approach. The remaining question, which specific design approach to use as a control condition, will be discussed in the next section.

5.2   Design approach used in control group sessions

In this section we explain which design approach we use in our control group session. There are several options for choosing the design approach of our control group session: 1) a custom-made session agenda per case, based on an intake, 2) developing a standard control group session, based on the average interests of all participating firms as expressed during the intake interviews, 3) developing an 'empty MuCh-QFD' session, that follows an MuCh-QFD format but that has its tasks removed that are effective in relation to the outcome quality requirements[13] (customer orientation to competitive positioning), and 4) use an alternative design method as control group session.

Before discussing these options in detail, we want to make some preliminary remarks. When creating a control group session there is a serious risk: participants must not suspect that they are in a control group setting. They need to feel that they are being tested equally seriously as the MuCh-QFD groups (or else differences between groups may be caused by differences in motivation and the experiment becomes invalid. Likewise, they need to get equal amounts of attention: participants' performance is likely to increase from the mere fact that somebody is watching. This is known as the Hawthorn effect). This means that the control group sessions have to meet the same requirements in terms of their internal logic and quality.

At the same time we know that differences between both session types have to be pronounced to generate statistically significant results. This has two reasons. Firstly this type of field experiment always has disturbing factors, which means that only pronounced effects can be measured. Secondly, this type of experiment with 'real world' design teams and business participants is time-consuming, and while finding around 30 participants for the experimental condition and 30 for the control condition is close to the practical maximum, it is also close to the statistical minimum, which again makes only the pronounced effects measurable. Thus, we

---

[13] As we explain below, performance on process quality requirements is more difficult to change, due to our choice to use a comparable session format, which creates many process similarities.

do not want the design approach of the control group session to be similar to MuCh-QFD or to function as an active trigger with regard to issues surrounding our design requirements. If we expect MuCh-QFD to generate high scores on the measurement instruments that we develop, we want a control group session that is likely to create low scores[14].

Thus, an important question regarding the set-up of our control group sessions is how to deal with the trade-off between high quality sessions and low performance on our design requirements. Below, we evaluate our four options. The evaluation is summarized in Table 5-2, plus and minus signs express the author's opinions. From this evaluation, we develop our final control group session.

Table 5-2 Evaluation of options to create content for control group sessions

	Methodology	Practicality	Reference Point
Custom-made agenda based on intake per case	(-) control condition is not constant (-) risks of MuCh-QFD bias in agenda due to focus of researchers	(+) feasible	(+) fitting for all cases (-) this is not the 'regular process'
Average control session agenda based on average interests	(+) control condition is constant (-) risks of MuCh-QFD resemblance	(- -) it is unrealistic to want to finish all intakes before first session	(-) what is the status of this 'average session'? Hard to relate to existing methods or theory (-) this is not the 'regular process'
'Empty MuCh-QFD', with MuCh-QFD formats, but effectiveness removed	(+) ability to check hypotheses on MuCh-QFD sub-tasks and effects (- -) low content sessions frustrate participants & results	(- -) very hard to facilitate (due to frustrated participants and lack of focus)	(-) what is the status of 'empty QFD'? (-) this is not the 'regular process'
Agenda based on other design method	(-) many methods have some overlap with MuCh-QFD (-) which method to choose – always somewhat arbitrary? (+) well-defined control session	(+) feasible	(+) compare MuCh-QFD with existing method (-) this is not the 'regular process'

Legend: (- -) = severe drawback; (-) = drawback; (+) = advantage

Potential approaches for defining control group session agendas are:
1. *Custom-made session agenda, based on intake*
   The main disadvantages are methodological. This approach will create

---

[14] This is similar to other experiments. If we want to investigate if watching anti-war movies increase pacifist attitudes more than normal movie watching, it is not wise to select control group movies that have a link with war or violence, since this may disturb the investigation.

biased agendas, polluted with MuCh-QFD elements, since at that point the moderators are engrained with the new MuCh-QFD method. A potential solution is to let somebody else prepare and facilitate the control group sessions, but this introduces a structural moderator-dependency to our experiment. Another disadvantage is that this approach makes the various control group sessions quite diverse. Hence, the ways that they differ from our experimental factor MuCh-QFD will not be constant. Because of its relatively high feasibility, this approach scores a plus on practicality. In terms of content an advantage is that the control session agendas will fit the cases. Nevertheless, even this custom-made session will not be the same as 'the regular service definition processes' that take place within firms, but this is a direct consequence of choosing a session format for control groups.

2. *Develop a standard control group session, based on average firm interests during intake*
The main problem is that this approach is unpractical: most firms will not want to wait long between the intake and the actual session and it takes at least three months to perform all the case intakes and devise a standard control group session. Secondly, in the area of methodology there is a risk that is similar to the one we mentioned with regard to the previous option: the compound desires plus our agenda suggestions may result in an 'MuCh-QFD-like' session, due to a strong focus on MuCh-QFD on the part of the moderators.

3. *Develop the control group session as an 'empty MuCh-QFD' session*
This is an approach where we use our MuCh-QFD method in an 'empty' way, by following the MuCh-QFD approaches, but removing all the elements that we expect to have effect. The main disadvantages of this approach are methodological and practical in nature. A session with so little sensible content will be dissatisfying to the participants (which invalidates the results) and moderators (thus making it practically impossible to facilitate a session like this for four hours). The advantage is that if we think we know what the main effects of the design activities on the agenda are, and if we make these hypotheses explicit, the 'empty MuCh-QFD' approach may be one of the most direct ways of testing the hypotheses.

4. *Choose one of the other service design methods as control method*
The main disadvantages are to be found in the areas of methodology and content. One methodology problem is that many design methods have some overlap with MuCh-QFD, which creates an impure control condition. Secondly, there is a certain arbitrariness in the decision which method to use. A methodological advantage is that the control condition is easy to control and can be linked to theory. Thus, an advantage of this approach is that it may result in conclusions regarding the performance of MuCh-QFD in relation to another known method. In terms of content a disadvantage is that the control session will not be the same as 'the regular service definition process' that takes place within firms, but this is the case for all four options.

Looking at the options described above, we decided first of all to discard options 1 and 2, since would both have generated intake-based agendas. Neither of these approaches is very reliable and, in addition, option 2 is too cumbersome in practical terms. Option 4 'other design method' and 3 'empty MuCh-QFD' could actually be seen as representing both ends of the trade-off described above (between high quality sessions and low performance on design requirements respectively). We decided to combine the two in a new synthesized approach. First we describe the steps that we have taken to 'strip' the effectiveness from the MuCh-QFD session agenda. Then we describe how we added a design method in such a way that we expected limited performance on the design requirements.

Table 5-3 Expected contributions of MuCh-QFD session activities on requirements

	1. Introduction (customer process, competitors etc)	2. Customer needs	3. Functions	4. Match functions & needs	5. Tasks of other channels	6. Competitors' benchmark
1. Customer Orientation	3	9		9		
2. Channel Coherence			3	3	9	
3. Channel Synergy			3		9	
4. Competitive Positioning	3	3		3		9
5. Progression	3	3	3	3	3	3
6. Focus	3	3	3	3	3	3
7. Stakeholder communication	3	3	3	3	3	3

Legend: 9 = strong contribution; 3 = moderate contribution – author's opinion

Table 5-3 contains the main activities performed in the MuCh-QFD approach. The scoring system used in this table again follows the scoring system of the QFD Relationship Matrix (Ramaswamy, 1996); a '9' indicates a strong contribution, where a '3' indicates a moderate contribution. If a cell is left empty, it implies that there is a limited effect from the corresponding instrument on the corresponding requirement. This table is based on our own judgments, and it is designed to help us in our 'empty MuCh-QFD approach'.

In our approach to generate an 'empty MuCh-QFD' session agenda the '9's are most important. These represent the greatest effects of design activities on requirements. One would want them to have a low impact in the 'control group' sessions. Consequently, the following steps were taken in the transformation towards 'control group' sessions:

1. For a low impact of the session on design requirement 1, customer orientation, the activities that score a '9' should not be predefined agenda topics. This does not mean a topic could not be discussed, but at least it is not defined in advance. In this case this means the following. In activity 1 of Table 5-3, the session introduction, we provided limited detail for the customer processes. Instead, customer processes were used as an illustration of the main phases (like pre-sales, sales and after-sales) and to indicate what the focal phase of the sessions would be. (This 'downsizing' of customer processes in the introduction was done in control group and

MuCh-QFD introductions to optimize experimental symmetry before the actual start of the design activities in the sessions. We did not consider this harmful to MuCh-QFD, since customer needs would remain an explicit starting point.) Table 5-3 activities 2, customer needs, and 4, match functions and needs, explicitly addressing customer needs, were removed from control group sessions.

2. To create a lower impact on design requirements 2, channel coherence, and 3, channel synergy, we removed activity 5, tasks of other channels, from control group sessions. Channel aspects that had been discussed during the intake, were also not used by the moderators for the introduction (activity 1) or other session activities.

3. There are two things that have impact on criterion 4, competitive positioning. Firstly, there is the intake. The intake addresses competitive positioning and influences are hard to eliminate here. The main decisions with regard to the position and definition of the e-service cannot be ignored (otherwise control groups would not be defining the same e-services as the MuCh-QFD groups). And we also decided to provide brief summaries of competing e-service initiatives in both session types. We are aware that this introduces a certain amount attention for competitors in the control group sessions. On the other hand, we wanted to use the summaries to make people aware that the success of new e-services should not be taken for granted [15]. Secondly, the competitors' benchmark, Table 5-3 activity 6, has a strong impact on the competitive positioning criterion. This activity is removed from control group sessions.

Eliminating activities is a useful approach to generate lower performance on the 'design content' requirements 1 to 4. However, with regard to the 'design process' requirements (progression, focus, stakeholder communication) this is not so simple. Design process performance depends more on the overall session approach (like using electronic group support systems, using an explicit intake, condensing discussions in short output focused agenda items, inviting different stakeholders, etc) than on specific activities. This is illustrated by the rows of '3's in Table 5-3: all activities contribute moderately to progress, focus and stakeholder communication, but there are no specific strong correlations. Our formats are to a large extent similar for both session types, and have to remain similar as a consequence of our choice of experiment (we want to test the effectiveness of MuCh-QFD tasks, not that of our session format versus other formats).

Above we have explained which e-service definition tasks we excluded on the basis of theoretical reflection (= transformation of agenda towards 'empty MuCh-QFD'). We now introduce the approach and activities that will once more provide 'substance' to the control group sessions (= transformation of the agenda towards another design method). We needed a formal design approach that was coherent and that, like QFD, followed structured steps, but that contained as little overlap as possible with MuCh-QFD. We found such a method in the product design

---

[15] This awareness should prevent participants from responding over-optimistically to the questionnaires; see also the pre-test results discussed in section 5.4.2, where we observed over-optimistic scores which we attributed to limited awareness of what it takes to make an e-service successful.

traditions discussed in chapter 3. The method we chose resembles traditional engineering as taught in technical schools and is known as 'traditional design', 'basic engineering' or 'partial design', see 3.1. We chose the name 'Fundamental Engineering' (FE) because is has few normative connotations, and is unlikely to bias session participants. The initial FE phases we describe in section 3.1 can be summarized by three key elements: a) objectives, b) functions and c) solutions. These three elements form the basis for our FE session development.

These three elements are translated into session agenda parts. FE sessions follow a progression through design objectives (Part I of FE sessions), functionalities (Part II of FE sessions) and solutions (Part III of FE sessions), see Table 5-5 for more details. The main exceptions to this structure are found in part III, where we ask participants to come with a service slogan summarizing the e-service proposition and to indicate whether they believe it would be worthwhile to invest in the proposition. The service slogan is used to provide FE groups with a stepping stone for starting their e-service design (our pre-tests showed that starting up this discussion was a big problem for groups). The reason we asked the participants to consider whether they would recommend investments, is to make them aware of how advanced an e-service definition needs to be to before investment decisions can be made (and thus to avoid over-optimistic responses to questionnaires). Asking whether or not people think a proposition warrants investment might lead to marketing discussions (and potentially raise customer, channel or competition issues), which could reduce the differences between MuCh-QFD and FE. We decided to incorporate these marketing questions in FE sessions anyway for the reasons we mentioned earlier and because they help being about an adequate flow in the FE sessions.

Table 5-4 Expected contributions of FE session activities on design requirements

	1. Introduction (Customer process, competitors etc)	2. Objectives	3. Functions	4. E-service Solutions	5. Investment discussion	6. Assess e-service in relation to objectives
1. Customer Orientation	3			3	3	
2. Channel Coherence			3	3		
3. Channel Synergy			3		3	
4. Competitive Positioning	3				3	
5. Progression	3	3	3	3	3	3
6. Focus	3	3	3	3	3	3
7. Stakeholder communication	3	3	3	3	3	3

Legend: 9 = strong contribution; 3 = moderate contribution – author's opinion

The fourth and final part of the FE agenda, evaluation of the e-service on design objectives, is meant to encourage participants to take a moment to reflect. It goes back to the starting point of the agenda: how well does the new e-service perform in terms of the objectives we formulated for ourselves? Besides the use this has for design purposes, this is also done for methodological reasons, to maintain symmetry between FE and MuCh-QFD sessions: it enhances people's perception that a session is well-rounded, which is likely to increase their level or satisfaction with regard to the session. Because a similar moment of reflection is built into the MuCh-QFD sessions, the same level of satisfaction can be expected. Because this element of satisfaction effect is not unique to MuCh-QFD, we wanted to include it in our FE sessions as well.

To conclude our control group session development, we listed the expected contributions of FE session activities on design requirements in Table 5-4. As can be observed, there are substantial differences between Table 5-4 and Table 5-3, and as far as FE is concerned there are only mild correlations (indicated by 3's) between agenda activities and requirements. This illustrates our expectations that FE session activities will contribute less than MuCh-QFD session activities to our design requirements, especially 1 to 4.

5.3 Description of 'Fundamental Engineering' control group session

The Fundamental Engineering (FE) session is used as a control group condition. In Table 5-5 the activities for the (FE) session are listed. In most session activities a GroupSystems™ group decision support system (GDSS) is used to generate and cluster input, and rank clusters via a vote.

Table 5-5 Fundamental Engineering session and process

Fundamental Engineering session agenda	Activities
Introduction	
- Present short review of the results of the intake	Presentation by facilitator
Part I: Objectives	
- Identify and cluster objectives	First GDSS brainstorm, then cluster via group discussion
- Prioritize objectives	GDSS vote, then discuss results
Part II: Functions	
- Define e-service functions for each objective	GDSS brainstorm
- Create e-service functionality clusters	Clustering via group discussion
- Prioritize e-service functionality clusters	GDSS vote, then discuss results
Part III: Solutions and new e-service	
- Define service slogan that summarizes proposition	Breakout exercise with flip-over
- Develop an e-service based on outcomes part I and II	Breakout exercise with flip-over
- Indicate why to invest or not	Breakout exercise with flip-over
Part IV: Assess the new e-service	
- Score the e-service on the objectives	GDSS vote
- Discuss IPA scores and whether objectives have been met	Group discussion

The introduction is identical to the one used in the agenda of the MuCh-QFD session. A brief summary of the intake result is presented by the moderator in the form of a presentation, with room for questions, discussions, etc. The presentation addresses the goal of the e-service definition session, a short background on our research project and procedures of the day, a short explanation of the design method that will be used, an outline of the e-service idea, why it would be valuable to customers, what the core product offer and the target segment are to which the e-service is connected, comparisons with competitive initiatives, and the expected financial benefits and costs. Part of the e-service idea explanation is to indicate which part of the sales cycle is chosen for the e-service definition session. This choice allows for speed and focus during the sessions.

Next, in part I of the agenda, the participants use e-service objectives as a starting point for the design process. First the participants brainstorm about the objectives, after which they cluster the objectives into categories. Finally, the categories are prioritized in a group vote using the decision support system.

Part II is aimed at defining e-service functionalities that would cover the objectives. For each of the clusters from part I functions are generated and then put into a central list. The resulting functions list is again clustered and prioritized, using group vote.

In part III the participants are taken away from their computers, to discuss three elements. First, the team summarizes the e-service proposition in the form of a service slogan, aided by the moderator. The service slogan is intended to create a shared image of the coherence between assumed customer value and marketing intention. This activity is partly inspired by the Service Concept Definition method as described in 3.1. The second element is an unfacilitated group exercise. The group receives their prioritized lists of objectives and e-service functions and is asked to work out the e-service solution in greater detail. This can be done either by drawing Internet pages or by listing and/or prioritizing the various elements that should make up the overall e-service. Finally, the moderator asks the team whether or not they feel the e-service warrants investment and why. Potential investment bottlenecks are listed on a central flip-over, and in some cases reasons for investments are added as well, insofar as they have been mentioned explicitly.

Part IV is meant to asses the newly defined e-service. It is used to reflect on the results, generate additional attention points that are important in the opinion of participants, and provide a psychologically useful feeling of 'closure' to the session. First, the electronic group vote is used to generate performance scores for the new e-service (i.e. How does the e-service score on objective 1? and 2? and 3? etc). These are then compared to the importance scores of part I, and the difference is calculated (= 'importance score' – 'performance score'), which generates so called 'Importance Performance Analysis' scores. These types of scores were introduced (Martilla and James, 1977) to assess the performance of professionals on certain attributes, and have been widely used since then to test the performance of services. At this point in the session, the moderator explicitly states that the scores generated in this session have limited external validity, but that they can provide a useful reference point for discussion. Then, the participants discuss the importance, performance and differences in scores with regard to issues that are surprising, confirm conclusions of previous discussions, or need further attention.

5.4   Pre-tests

In the previous sections we described the MuCh-QFD and FE sessions. However, we did not arrive at these final formats at once. In this section we describe the pre-tests that we performed, and the choices we made based on their results.

We pre-tested our e-service definition sessions in three steps from February to September 2004. In the first pre-test we tested an initial version on two groups of managers from a large insurance firm. In the second pre-test we tested a more standardized version of MuCh-QFD, using an electronic group decision support system (GDSS) of GroupSystems™. We also tested it against 'Fundamental Engineering' control group sessions. As a consequence, we made several small modifications to both sessions. In the third pre-test these modifications were tested again, and left unchanged. This means that the sessions acquired their final form in the tests we conducted in a 'live' environment.

5.4.1   First pre-test

We pre-tested a first version (without GroupSystems™ support) of our MuCh-QFD session with a large financial firm (banking and insurance services) in February 2004. Two parallel groups, of four and three participants respectively, used an early version of our MuCh-QFD method for e-service design in a multi-channel context. The groups were multi-disciplinary, with participants from operations, marketing, finance and ICT. The intake phase of our method was performed via a preceding session with about 15 firm representatives with varying backgrounds, who were asked to formulate three e-service possibilities. From these, an e-service was chosen that provides online support for mortgages based on mutual funds. This case was considered suitable because of the high level of support customers require and because of the importance of intermediaries in servicing customers (multi-channel aspects). The pre-test was evaluated in two ways. Individual feedback was collected by means of a questionnaire aimed at gathering opinions about the positive and negative aspects of the session, and judgments regarding the impact of our design method on the seven design requirements. Also, we collected feedback via group discussion.

First pre-test results and consequences for design of the field experiment
The main finding of the pre-test was that the session tasks were neither concise nor precise enough. There was too much room for different interpretations of the session tasks. As a result, there were significant differences between the two groups in terms of the design process and of the results, despite the fact that they followed the same MuCh-QFD method. For example, group A described the customer process in much greater detail than group B, and used less detail for functionality for each of the process steps. All this was confirmed in the individual and group evaluations. The participants complained that descriptions of the tasks and design process lacked precision and suggested we provide more examples of the output that had to be generated in each step. They also complained that few

decisions were made. In general, opinions varied or were neutral regarding the sessions' contributions to channel synergy, channel coherence and competitive positioning. Although they did see some potential, they said there was enough progress. In the individual as well as the group evaluation participants were most positive about the level of customer orientation and about speed and stakeholder communication. Opinions varied with regard to design process focus: group A was clearly less positive than group B. As a consequence of these findings, we defined our design tasks more narrowly, we created input examples, and we decided to use a group decision support system (GDSS) to manage the design process.

Moreover, there were three additional observations that we translated into consequences for the design of our field experiment. Firstly, the sessions diverged too much (collecting information via structured brainstorms), and converged too little. In other words, the groups discussed several issues, generated a large amount of output, but on a psychological level failed to reach 'closure'. This is why we changed our protocols in such a way that each idea generation phase was followed by prioritization: 'which customer needs, objectives, functions, solutions, etc are most important?' Secondly, we saw that the scope of our sessions was still too broad, so we decided to precede sessions with a formal intake, using a fixed intake protocol. This protocol covers several e-service definition tasks and helps create focus for the subsequent sessions, as also explained in 4.2. Thirdly, the effect of the moderator appeared to be strong. Group B had more sense of direction and more focus than group A, and used different approaches to address e-service definition tasks, which appeared to be caused by a combination of too much freedom in session protocols and compensation by tacit knowledge of moderator B. Therefore, in addition to introducing stricter process management with the GroupSystems™ decision support system, we decided to check in our second pre-test if our modifications had made our sessions more 'moderator-independent'.

5.4.2 Second pre-test

In our second pre-test we tested the effects of our MuCh-QFD sessions against control groups that followed the 'Fundamental Engineering' approach. Following our first pre-test results we decided to give our second pre-test a rather strict experimental design. Our primary research question was:

> RQ1: Do our MuCh-QFD sessions improve the design process for our seven design requirements in comparison to our 'Fundamental Engineering' control sessions?

Our propositions were that we would find higher MuCh-QFD scores for the quality-related requirements of customer orientation, channel coherence, channel synergy and competitive position, and for the three process quality requirements of speed, focus and stakeholder communication. We expected it would be harder to show the process differences experimentally, since the control sessions have a number of process characteristics in common with the MuCh-QFD sessions.

Our second experimental factor addressed the influence of the session moderator. We realize there is extensive literature on this topic. We do not discuss this literature since this is outside the primary focus of our research. But on a practical level, we did want to check the impact of the session moderator in our experimental design of the pre-test. Hence, our second research question is:

RQ2: Does a different moderator lead to different performance on our seven design requirements?

Our MuCh-QFD and FE sessions were developed to be effective regardless of who is the moderator, after such a moderator has had sufficient training. Our proposition was that we would not find significant differences between moderators on any of our seven requirements.

2x2 Experiment

Our second pre-test was performed on April 2004 with master's students with a design background, following the experimental design of Table 5-6. Each group was given the same assignment. They were to act as a design team of a mobile phone manufacturer, and design a new e-service. They were all well acquainted with the Internet sites of mobile phone manufacturers (as we checked via background questions in our questionnaire). Each group received the same background information, which was presented to them as the result of the intake phase of our approach. They had to fulfill the next phase of our approach: the e-service definition session. Five participants were scheduled for each session, but due to last minute cancellations, on average only four students participated. For reasons availability we chose students as participants. Thanks to their background the e-service definition task we gave them was not alien to them. The external validity with regard to problems firms face in their real environment remained a question, however, because the assignment involved an artificial problem and the students had no experience with corporate service development. Our second pre-test was more a structured case study than a controlled experiment, since there were disturbing factors in the form of potential differences in participant backgrounds and group effects (see also below). Also, there were too few participants to use robust statistical tests like ANOVA analyses. Despite these limitations, our research design enabled us to test the same case assignment with relatively similar participants across 2x2 conditions.

Table 5-6 Experimental design: MuCh-QFD versus FE, plus moderator influences

	MuCh-QFD Session	FE Session
Moderator 1	Session 1 (4 participants)	Session 2 (3 participants)
Moderator 2	Session 3 (4 participants)	Session 4 (5 participants)

Measuring performance

We measured performance in two ways. We asked subjective opinions with regard to usefulness on a 5-point Likert scale (very useful to very useless) regarding our seven requirements (e.g. 'Do you think the session is useful to make the service design customer centric?' etc). And for each design requirement we asked whether according to the participants certain 'objective' events had occurred (between 6 and 9 items per requirement). For example, regarding customer centricity, the first two items were: 'Did the team discuss what the five main customer needs are?' and 'Have customer needs been prioritized?' These items were used as additive scales, generating maximum scores of 6 to 9 points per requirement (for details see appendix E). However, since additivity of the data was not proved, for safety we used Mann Whitney, or Wilcoxon, nonparametric tests for comparing effects of different experimental conditions (and as a cross check: t-tests generated the same significance levels).

Second pre-test results

We have several types of results: our measurements, our own observations and the output of the groups. As far as our two types of measurements are concerned, the subjective opinions regarding usefulness did not generate any significant differences between our experimental conditions. This why from here on we focus on the objective event measurements.

The number of participants was not high enough to analyze variance across both variables and test for interdependence. Hence, we performed a separate Mann-Whitney test for each condition. Regarding our second research question - 'is there moderator-impact?' - our proposition was that we would not find any differences. There was a significant difference, however, for one requirement: $p < 0.05$ for channel synergy. Based on our own observations, we believe this may well be attributable to group effects. We address this further in our pre-test discussion.

Regarding our first research question, 'does our MuCh-QFD session improve the design process?', we observed that the averages were higher than for FE sessions on all seven requirements. However, only three differences were significant, based on Mann Whitney tests: customer orientation ($p < 0.005$), competitive positioning ($p = 0.01$) and communication ($p < 0.05$). These levels of significance are striking, given the small number of observations ($n = 16$).

Our own observations have led us to believe that a) MuCh-QFD performed better on these three requirements: customer centricity, competitive position and stakeholder communication, that b) measurements on channel coherence and speed could become significant with a larger number of participants, but that c) using the measurements on channel synergy and focus as differentiators would be cumbersome. Regarding synergy, several items were not discriminatory; e.g. currently most click & mortar e-services are likely to 're-use existing brand value'. Also, several items needed to be rephrased to create a greater focus on multi-channeling and/or on design process observations. With regard to focus, it proved to be hard to make discriminatory measurements. In all sessions the scores were exceptionally high. For both methods and both moderators the averages were 7.0 to 7.4 out of 8 items. We checked the data, and the items that showed no discrimination between the methods were removed. Also, some items were reformulated. Regarding our aim of supporting a focused design process, everybody agreed that the sessions, using GroupSystems™ support, were highly focused. So the good news is that in terms of focus our e-service definition sessions appear to be successful. However, this finding is not proven via robust control group / experimental group testing.

In our FE pre-test sessions, one of the final agenda activities turned out to have limited added value. Originally, in part IV of an FE session -evaluating the priorities for the objectives of the e-service, which had been set during part I of the FE session were evaluated. However, the participants did not see the value of this activity and they said that it confused them, since it was too similar to the previous activity on the agenda (which was to indicate how well they thought the new e-service scored on the design goals the participants had defined in part I). This is why we decided to remove the activity from the agenda.

The version of MuCh-QFD we used in our second pre-test contained a brief cost estimation. This did not yield very useful discussions or outputs, so we decided to exclude it from the session after the second pre-test. Instead, we present a high-level business case in the intake and postpone other cost discussions until after the session, when more design details are known. Finally, participants said that as far as they were concerned, the last item on the agenda – which originally was value engineering –the least useful item of our second pre-test version of MuCh-QFD. The item addressed ways to optimize customer value or reduce costs. Because of participant feedback we decided to remove this item from the MuCh-QFD agenda after the second pre-test. However, there are two things we want to emphasize with regard to value engineering. Firstly, we think that FE participants would have judged differently if their last item had been value engineering. This is the case because in their sessions less attention was paid to customer value or supplier value. Secondly, the comments of the participants of the MuCh-QFD session appear to be sensible. They had already optimized their design to quite an extent in relation to customer and supplier value in the previous design tasks. Hence, without additional information or insights, the final item regarding value engineering added little. Indirectly, this confirms the value of the other MuCh-QFD design tasks.

Discussion
We observed several group effects. We think the 'moderator impact' in the channel synergy scores we found is in fact a 'moderator artifact', which is at least partly caused by the combination of singularities in groups 1 and 4. According to our observations, group 1 was focused and competent, generating relatively high quality discussions and decisions. At the same time, they were most critical regarding their objective scores. We think the 'moderator artifact' in the channel synergy scores we found is partly caused by their strictness: 'This synergy may be present, but we did not discuss it, no'. If they had been less strict, like the other groups, they would have scored higher. As indicated earlier, particularly the items under 'channel synergy' need to be rephrased to capture the design process better, and appear most sensitive to different interpretations. Group 4 appears to be another group responsible for the channel synergy moderator artifact, since they scored relatively high in the questionnaire. This group developed a line of argumentation that was directed mostly at manufacturer interests. This had two effects. Firstly, they appeared to be very satisfied with this, creating somewhat optimistic interpretations of all the items (e.g. their scores on customer orientation and competitive position were higher than those of group 2, which was the opposite of what one would expect in light of the fact that they paid less attention to these items). We discuss this 'satisfaction' effect below. Secondly, their supplier focus did create more synergies, which elevated their scores. Together, the diminished scores of group 1 and the elevated scores of group 4 could explain the channel synergy moderator artifact.

Session 2 had a group effect in terms of creativity and graphic orientation. During the breakout, they immediately went for the drawing board and started sketching Internet pages. This led to a positive and rapid flow of ideas and few dissenting opinions. However, this does not seem to have disturbed our measurements very much. Group 3 was most different from the other groups: in terms of participant backgrounds and of their behavior as a group. Three of the four participants had cultural and national backgrounds that were different from those of the rest of the 16 participants. This made communication more difficult. Also, two of the participants in group 3 were reluctant to express their ideas and opinions. This meant that it took longer for ideas to be translated into opinions. We observed that the strict MuCh-QFD process helped them 'pull through' and create focus and direction. We believe that the results would have been less satisfactory had this group been subjected to control group conditions.

Table 5-7 Channel synergy 'moderator artifact' due to group effects sessions 1 & 4

	MuCh-QFD Session Avg Participant Score	FE Session Avg Participant Score	Average MuCh-QFD & FE
Moderator 1	4.8 ('too low', group 1)	4.0 (group 2)	4.4 ('too low')
Moderator 2	6.5 (group 3)	5.8 ('too high', group 4)	6.1 ('too high')
Average Moderator 1, 2	5.6	5.1	5.4

Another disturbing factor was the limited experience the students had with corporate service development. This factor disturbed our results in two ways: via an 'awareness effect' and a 'satisfaction effect'. With the 'awareness effect' we mean that the MuCh-QFD groups became more aware of design complications (regarding multi-channel considerations, uncertainties in costs and revenues), since addressing these complications was an integral part of their sessions. This made them interpret our 'objective' questions more strictly (e.g. 'We did talk about other channels, but did not really consider all their (dis)advantages'). Our control groups, on the other hand, appeared to be more easily satisfied with their thoroughness after addressing issues besides Internet functions and solutions (e.g. 'Yes, one of us mentioned call centers'). The 'satisfaction effect' that we observed in MuCh-QFD and FE groups was that participants were impressed with what they could achieve in one afternoon. This positive attitude appeared to increase their scores on the various items (e.g. 'Yes, we talked about this and we mentioned that too.') Hence, the differences between 'we really discussed XYZ' and 'XYZ was mentioned once' become smaller, and the discriminatory effects of our items are reduced. We also believe that the 'awareness' and 'satisfaction' effects had an impact on our subjective measurements, which turned out to be entirely non-discriminatory.

As we expected from design theory (Cross, 1994) an initial overview of customer needs is seldom exhaustive. When extending the design space to channels other than the Internet, other needs and other functions are identified. These needs are important for the overall service design, which confirms the value of our additional design step to address the functions of other channels.

Summary of consequences for MuCh-QFD and FE sessions

To summarize, we translated the second pre-test results into the following consequences:

- The majority of MuCh-QFD and FE agenda items were accepted, as well as the use of GroupSystems™ to manage the design process.
- We removed the re-evaluation of the priorities of design objectives from our FE agenda, since it appeared to add little and sometimes even confuse participants.
- We also removed value engineering and a brief cost discussion from our MuCh-QFD agenda, since they also appeared to add little.

5.4.3 Third pre-test

Our third pre-test was conducted in September 2004. We used two student groups for two sessions. The students had alternating roles: either they were participant in the first session and observer in the second session, or the other way around. First, we conducted one MuCh-QFD session with six participants and five observers. Secondly, we conducted one FE session with five participants and six observers. Due to practical limitations, this pre-test was conducted on a small scale. The advantage of the design of this pre-test was that the participants were able to compare and provide improvement suggestions with regard to both types of session.

During the third pre-test, the main thing to do was check if the flow of the new MuCh-QFD and FE agendas was logical to us and to the participants. As explained in section 5.4.2, the MuCh-QFD agenda had undergone several small changes between pre-test two and three. Value engineering and a brief cost discussion were removed from the MuCh-QFD agenda, because they added little value. We also moved several agenda activities that addressed competitive position from part II to part IV of the MuCh-QFD agenda. The FE session was also slightly modified before the third pre-test: as explained in section 5.4.2, the re-prioritization of design goals was removed from part IV of the FE agenda. During the third pre-test we found no further problems or weaknesses in our session agendas, so we decided to leave them unchanged. With the MuCh-QFD and FE agendas now fixed, we hypothesize below on the differences we expect to encounter regarding their impact on the seven design requirements.

5.5 Hypotheses on MuCh-QFD and FE differences

In this section we formulate the hypotheses regarding the differences between MuCh-QFD and FE session results, which we attempt to validate in our field experiment. Our hypotheses are strongly linked to the agenda parts of the sessions. We start with hypotheses for each part of the agenda, and then address hypotheses on the overall differences. The hypotheses we formulated are relatively common sense. This is a consequence of the fact that in design theory it is not common practice to test these types of hypotheses empirically with regard to design task effectiveness (see also our discussion in section 1.3): thus we start testing hypotheses 'from the ground up', and use this research as a first step to test some of the most straightforward assumptions.

We begin by addressing the design *outcome* quality related requirements 1 to 4 (customer orientation, channel coherence, channel synergy and competitive positioning). An initial remark is that we expect a relatively peaked pattern for the MuCh-QFD sessions, with a strong emphasis on only one or two of the design content related requirements for each of the agenda parts. See also Table 5-8.

Table 5-8 Relation MuCh-QFD session agenda, QFD 'House Of Quality' and design requirements

MuCh-QFD session agenda	Connection to QFD HOQ or QFD-extensions	Design requirement
Part I: Customer needs and Internet functions	Room 1 & 2	Customer orientation (1)
Part II: Define e-service	Room 3	Customer orientation (1)
Part III: Tasks of other channels	Multi-channel (room 3)	Channel coherence (2) & channel synergy (3)
Part IV: Competitive position	Room 4	Competitive position (4)

With regard to the FE sessions, we expect a much less peaked pattern, since customer orientation, channel coherence, channel synergy and competitive positioning have not been identified as explicit attention points in specific parts of the agenda. This, together with Table 5-8, leads to the following hypotheses for each of the agenda parts:

H Part1.Req1
Agenda part I shows a significantly higher level of customer orientation for MuCh-QFD than for FE sessions

H Part2.Req1
Agenda part II shows a significantly higher level of customer orientation for MuCh-QFD than for FE sessions

H Part3.Req2
Agenda part III shows a significantly higher level of channel coherence for MuCh-QFD than for FE sessions

H Part3.Req3
Agenda part III shows a significantly higher level of channel synergy for MuCh-QFD than for FE sessions

H Part4.Req4
Agenda part IV shows a significantly higher level of competitive positioning for MuCh-QFD than for FE sessions

It is possible that competitive positioning comes up in part III of FE sessions, as a result of the investment-related question, which may lead to discussions about obtaining a competitive position. However, on average we do not expect this effect to be very pronounced, which is why in this case we do not present a separate hypothesis.

Also, several general hypotheses can be formulated. Following the previous explanation that no explicit attention to customer, multi-channel or competition issues was built into the FE sessions, we expect the overall scores to be lower for FE sessions. Based on this assumption, our overall 'design content' hypotheses are:

> *H Req1*
> *MuCh-QFD sessions show a significantly higher level of customer orientation than FE sessions.*
>
> *H Req2*
> *MuCh-QFD sessions show a significantly higher level of channel coherence than FE sessions.*
>
> *H Req3*
> *MuCh-QFD sessions show a significantly higher level of channel synergy than FE sessions.*
>
> *H Req4*
> *MuCh-QFD sessions show a significantly higher level of competitive positioning than FE sessions.*

Secondly, we address the design *process* quality related design requirements for each of the agenda parts. For these requirements we do not expect clearly pronounced peaks for specific agenda items. Both session types use a format that has been designed to be efficient and fast (= requirement 5, 'speed' or 'amount of progress'), focused by the use of the electronic group support system which provides a high degree of structure (= requirement 6), and facilitates the exchange between multiple stakeholders (= requirement 7), due to explicit selection of different participants and due to the 'democratic' nature of sessions thanks to the electronic group support systems (inputs and priorities are generated equally, and hence not dominated by one or two people, as is often the case in group sessions). Apart from our expectation that patterns will not contain many peaks, we also expect to find a limited number of differences between the MuCh-QFD and FE sessions regarding the design process. This is illustrated in Table 5-9, in which the differences and similarities between the sessions are summarized. Nevertheless, we expect there will be some minor differences.

In general, we expect the differences will often be too small to become significant in terms of progress and focus (requirements 5 & 6). But in relation to agenda part III of FE sessions, which is less structured than other tasks and has a relatively ill-defined goal of determining what the e-service will look like, we expect that there may be less focus than part III of MuCh-QFD sessions:

> *H Part3.Req6*
> *Agenda part III shows significantly less focus for FE than for MuCh-QFD sessions*

Table 5-9 Differences and similarities between MuCh-QFD and FE sessions

	MuCh-QFD sessions	FE sessions
Intake review	Yes	Yes
Session format (electronic/speed/focus)	Yes	Yes
Customer needs as starting point for functions and for design evaluations	Yes	No
Multi-channel coherence and synergy considerations	Yes	No
Competitors' benchmark	Yes	No

Regarding requirement 7 - communication between multiple stakeholder perspectives - we expect an increasing number of stakeholder perspectives to be addressed during MuCh-QFD sessions. We do not expect the same increments during the FE sessions, since the different perspectives of customers, other channel participants, supplier desires in relation to competitive positioning are not addressed as structurally as in MuCh-QFD sessions. Hence, we expect a) the scores on requirement 7 to increase from part I through to part IV of the agenda, and b) the sum of scores on the various agenda parts for MuCh-QFD sessions to be higher than for FE sessions, because the customer, channel and competition issues are explicitly addressed in the various MuCh-QFD agenda parts. As a result, our hypotheses are:

H Part1/4MuCh-QFD.Req7
Agenda part IV shows a significantly higher level of stakeholder communication than agenda part I of MuCh-QFD sessions.

H SumofParts.Req7
The sum of agenda parts I to IV shows a significantly higher level of stakeholder communication for MuCh-QFD than for FE sessions.

Finally, as far as the general process hypotheses on sessions are concerned, we do not expect significant differences. This is due to similarities in the MuCh-QFD and FE sessions, with the possible exception of requirement 7, for the reasons we just mentioned. The corresponding hypothesis to be tested is:

H Req7
MuCh-QFD sessions show a significantly higher level of stakeholder communication than FE sessions.

We return to these hypotheses in chapter 7, where we discuss our empirical results.

5.6 Conclusion

In this chapter we determined which e-service definition process to use as a control group condition. Our first decision was to choose a session format for the control groups (instead of the 'regular' e-service definition processes taking place in firms, for example), for methodological and practical reasons. The main methodological advantage is that if we compare MuCh-QFD sessions to sessions that are only different in terms of their content, namely which design tasks are performed, we have created conditions that are sufficiently well-specified to be able to compare the effects of design tasks quite precisely. The design content of control group sessions is based on 'Fundamental Engineering' (FE), which resembles the kind of traditional engineering that is taught in technical schools. We made comparisons between MuCh-QFD and FE for each agenda part and generated hypotheses on the differences between design task effectiveness in relation to the seven design requirements.

A handicap is the fact that a control group session creates an 'artificial' reference point for our comparison with MuCh-QFD. Control group sessions can be expected to have greater focus and progress than 'regular' e-service definition processes. They also involve the main stakeholders, although cohesion between stakeholder perspectives is stimulated less than in MuCh-QFD sessions. It is therefore doubtful whether MuCh-QFD will perform better than control group sessions with regard to the design process quality requirements. However, if MuCh-QFD scores are higher, they can also be expected to be higher than the scores that 'regular' processes would have obtained. Regarding design outcome quality requirements (customer orientation, channel coherence, channel synergy and competitive positioning) more pronounced differences are expected between MuCh-QFD and FE, as illustrated by our hypotheses on the effectiveness of each of the design tasks. If we find different scores between MuCh-QFD and FE, these can likely be attributed to specific design tasks. On the one hand a drawback of this approach is that it it is not possible to determine directly how these scores relate to the degree of difference between MuCh-QFD and the 'regular' e-service definition processes in firms. On the other hand, FE sessions provide ample space for participants to determine their own objectives and e-service definition priorities. If, for example, half of the FE sessions show limited customer orientation, that does support the conclusion that design teams, when left to themselves in determining the regarding the priorities in e-service definition, run the risk of neglecting customer orientation, in comparison with MuCh-QFD.

6   Research Design

If you want to become perfect at something, find something that you can repeat (Anonymous)

Case Exhibit 6-1 Illustration of testing the observation protocol in the third pre-test

Our third pre-test was conducted in September 2004. We used two student groups for two sessions. The students had alternating roles: either they were participant in the first session and observer in the second session, or the other way around. First, we conducted one MuCh-QFD session with six participants and five observers. Second, we conducted one FE session with five participants and six observers. One of the goals of the third pre-test was testing the newly developed observation protocol:

The observation protocol consisted of a sheet with yes/no items for each design requirement. Observers had to cross a yes box when they encountered the element in question on the observation sheet. In total, there were 37 yes boxes, divided across seven categories. At the end of an agenda activity, they had to total the yes-scores per requirement to a score per requirement, and then take a new sheet for the next agenda activity. During the third pre-test, the observations were performed at a very detailed level: the MuCh-QFD agenda consisted of 18 activities, which meant that 18 observation sheets had to be scored. Observers complained about this level of detail: they felt it was cumbersome, and we feared this would reduce measurement reliability. Also, from our experimental point of view, we concluded that a higher level of abstraction, clustering and measuring in four main agenda parts per session, made more sense. It followed the internal logic of the sessions more accurately. Hence, we decided to use our observation protocol to generate four measurement episodes per session.

In this chapter we answer our fifth research question:

5. How can we evaluate the performance of our (experimental and control group) e-service definition methods with regard to the design support requirements?

In this chapter we describe our research design in section 6.1, and in section 6.2 the field experiment that we conducted. Then we describe development of our two participant and observer based measurement instruments in section 6.3. In section 6.4 we discuss reliability and validity of those instruments.

6.1 Causal model and research design

In this section we first explain the general causal model and research design. Then we discuss the nature of our empirical test, i.e. a structured field experiment. After that, we address the scope of the measurements we will conduct, and how and why we accept a certain degree of bias in our measurements. Finally we describe the number of cases and participants for each experimental condition.

6.1.1 Causal model and experimental factor

The basic causal relationship to be investigated is:

X_e (Design Method) $\rightarrow$ O (e-Service Definition Quality)

The experimental factor X_e is the design method used and the effects, O, are measured in terms of design requirements (customer orientation, channel coherence etc). Thus, the causal model underlying our field experiment is rather simple. Because we want our method to be relevant to real-life e-service definition problems, we choose to test the impact of our method on real world e-service innovation initiatives, with real business participants in our sessions.

To investigate this causal relationship we chose to work with a research design that is very similar to the research design called 'static group comparison' (Hagenaars and Segers, 1980). This means that we use an experimental group versus a control group. And as explained in chapter 5, we use a variation on 'the classical experiment' (where there is an absolute absence of the experimental factor in the control groups) in the sense that there is actually a different design method presented to control groups instead of none. To ensure a maximum degree of comparability between the two groups, we randomly split each design team into two groups: half of them attended our MuCh-QFD session and the other half attended the control group FE session. In addition to collecting data via questionnaires and interviews with participants at the end of sessions, we also perform measurements during all parts of sessions, using external observers.

In standard notation (Campbell, 1957) our research design is:

R X_e + O1   O2
R X_c + O3   O4

The experimental factor - X_e - is the MuCh-QFD session. The control factor, X_c, is the FE control group session we developed (section 5.2). In the notation above, R stands for the randomization of participants across both conditions, as explained above. O1 & O3 reflect for observations during the design sessions. Each design session was divided into four main parts (in addition to an introductory part 0 of the agenda and a questionnaire at the end), and for each part (I to IV) a separate measurement was performed by the observers. Hence O1 & O3 actually consist of four consecutive measurements. O2 & O4 stand for participant questionnaires at the end of the design sessions. To gain additional insights and also to ensure a correct response to the questions, all participants were interviewed on the basis of the answers they provided in the questionnaire, immediately after they completed their questionnaires.

Factors that could have an influence on the basic causal relationship between design method and impact on the design requirements (such as firm size, maturity of the e-service idea, established design practices of participating firms etc) were not included in our research design as experimental factors, because this would have resulted in too few observations for each experimental condition. With our experimental design, dividing the participants between the two experimental conditions with each case, we tried to control for these factors (to the extent to which this is possible in a field experiment approach like this). We also collected data on participant characteristics (like number of years in the industry, function, and previous experience on design, service innovation or sessions), which we used to analyze disturbing influences.

6.1.2   Nature of empirical approach: 'structured field experiment'

Our aim was to obtain at least 30 observations for each experimental condition (MuCh-QFD versus FE). However, even with 30 observations per experimental condition, our field experiment is not very well-controlled with regard to case-dependent conditions. This is not so much a matter of statistics, but more of methodology. Firstly, although group dynamics are controlled to some extent by the high degree of structure we provided, we do want groups to be able to discuss issues freely, and this introduces differences between the sessions. This means we are quite vulnerable to group effects (for example a session might be dominated by one person, or the degree of 'chemistry' between participants may vary per session). Secondly, each single case can have different background conditions (e.g. the type of e-service chosen, participants' background, maybe company politics or previous experiences in the background that influence the discussions, etc). Hence, potentially there are many disturbing factors present, which we cannot control or compensate for. This also means that the participants in each of the two experimental conditions do not operate entirely independently. The extent of independence is a factor to consider in the interpretation of our results. In fact, interpretation will be an important element in our analyses: the numbers only tell part of the story. This is why we have decided to collect qualitative case data as well, see also section 6.2. Since our research design is positioned between a controlled experiment and a case study approach, we would like to label it as a structured field experiment.

6.1.3   Nature of our measurements

As stated in section 6.1.1 we used measurement instruments for our core concepts. There are two points we want to address in this section: a) the scope of our measurements, and b) our decision to allow a certain bias in our measurements. The development of the measurement instruments themselves is described in section 6.3.

The first point we address is the scope of our measurements. We want to measure the effectiveness of the MuCh-QFD session in relation to a number of design requirements. However, we only measure certain aspects. Firstly, our method is only aimed at the initial service definition phases. The result of a session is not a final design, nor does it include an implementation. This means we cannot use quality of implementation or commercial success as a yardstick (And even with market introduction and commercial success present, it would remain practically very difficult to attribute commercial success to a design method, without very large numbers of market introduction, since disturbing factors play such a significant role here). Hence, we limit ourselves to the service definition phases. Secondly, we have chosen to measure design process instead of service definition results. In our opinion, though they are desirable, quantitative outcome measurements for e-service definitions are hard to obtain in practice. One disadvantage is the small number of observations. For us, with six cases, we would have six observations for each design condition. In view of our resources, performing five times as many sessions (to generate 30 observations per experimental condition) was not practical. Another disadvantage is the difficulty to generate objectified and reproducible results. To our knowledge very little work has been done on how to measure the quality of a service definition (and we think it is a very challenging task to define this in relation to, for example, the degree of customer orientation, the degree of channel coherence, or the degree of marketing fit and competitiveness). Hence, we focus on design process measurements: 'what has occurred and how customer-oriented etc have the discussions and decisions been?' One point of clarification to avoid misunderstanding: in previous chapters we distinguished between content (or 'outcome') quality-related requirements and 'process' quality-related requirements. It is possible to measure both types of requirements from a design process point of view. It simply means that we do not attempt to measure whether the *outputs* were customer-oriented etc, but the *degree of customer orientation of the discussions and decisions during the e-service definition process*. Hence, our instruments are based on the judgments of people (participants and observers) with regard to the degree of attention, discussion and relevance attributed to aspects like customer centricity, channel coherence etc.

As the second topic of this section, it is good to clarify that there is a certain bias in our measurements. This bias was introduced knowingly and warrants an explanation. There are two forms of bias. Firstly, we only used constructs for the design requirements for which our MuCh-QFD session was optimized. Secondly, discussion topics that were not planned in our sessions, e.g. usability, financial or technical discussions, did not generate measurement scores, since they were not included in our item lists. Some of those elements (e.g. finance, user-friendliness) were originally present in our constructs, but removed because they were not structural elements of our final session agendas.

We acknowledge that our measurement bias has the disadvantage that in all likelihood we have not captured all valuable discussion elements in the sessions. However, this was also not our purpose. Our purpose was to test whether a MuCh-QFD session is relatively effective in supporting the design requirements we aimed to support. More precisely, we expected there would be effects at specific parts of our session agendas (see section 5.5 for our hypotheses). From this perspective we consider the biases mentioned above a logical consequence of our aim: measuring whether the MuCh-QFD e-service definition tasks generate the effects that we hypothesized on the requirements customer orientation to stakeholder communication.

6.1.4 Number of participants per experimental condition

After having performed eight pre-test sessions (see section 5.4), our final experiment consisted of 12 e-service definition sessions with business participants. These sessions came from six cases, where each half of a case team was exposed to either the MuCh-QFD or FE method.

Table 6-1 Number of participating cases and participants

Cases	Sector	MuCh-QFD Participants	FE Participants
Case A – Oct 4th 2004	Insurance	4	4
Video Communication Case	Insurance	canceled	Canceled
Case B – Oct 13th 2004	Insurance	4	4
Savings Plan Case	Insurance	canceled	Canceled
Case C – Oct 19th 2004	Insurance	4	4
Case D – Dec 2nd 2004	Insurance	4	4
ISP Services Case	Telecom	canceled	Canceled
Case E – Dec 7th 2004	Telecom	4	5
Case F – Dec 8th 2004	Telecom	4	4
Consumer Billing Case	Telecom	canceled	Canceled
Total		24	25

In Table 6-1 the number of participants per condition is shown. We performed two runs to collect cases and conduct sessions. Hence we had 'October sessions' and 'December sessions.' Originally, our aim was to use four more cases, to get more than 30 participants for each of the two design conditions. From a statistical point of view, the number of 30 respondents per experimental condition is often mentioned as a minimum for realizing solid statistics. Unfortunately, due to practical limitations we had to make do with six cases. (Although the cases are from two industries, insurance and telecommunication, a priori we do not consider this a separate experimental condition. To make sure, we check for differences between the two industries in section 7.2.) Two of the main limits we encountered were time and the availability of cases and participants. To start with the latter, initially it looked as though we had six cases in October. However, one of them turned out to be not that suitable for our test and it was difficult to find the participants for. Another case withdrew because the company's internal priorities changed. A third case had to be delayed until December, because it proved to be very hard to arrange the participants. For the remaining three October cases, a lot of time was spent obtaining participants for each session who were able and willing to represent the stakeholder perspectives we needed. Also, in the December-run two cases with varying potential did not make it in the end. Finally, our time-constraints meant that the projects in which we could collect cases and conduct sessions ended by December 2004.

Despite the fact that we did not reach 30 participants for each of the experimental conditions (MuCh-QFD versus FE), and we agree that this limits our possibilities for quantitative analyses, we decided to conduct quantitative analyses anyway. In terms of the physicist: 'The difference between 25 and 30 participants is only 20%, which means that the signal to noise ratio for 25 participants is only 10% lower than for 30 participants (which equals the difference between explained and unexplained variance). Thus, if a difference is significant at the 0.01 level for 25 participants, it is unlikely to become not significant at the 0.05 level with 30 participants.' This is the numerical reason why we decided to proceed with this data to perform our quantitative analyses as planned, with some additional caution.

Secondly, there is a methodological reason, from the field of user tests with new designs, to perform analyses on the basis of less than 30 observations. Unlike the social sciences and mathematics, where the convention is to use 30 participants per condition, there are other communities of practice within usability engineering and industrial design that prescribe much lower numbers: roughly five to eight observations per experimental condition for evaluation studies (Nielsen and Landauer, 1993; Beyer and Holtzblatt, 1998; Nielsen, 2004a, 2004b; Norman, 2004; Tullis and Wood, 2004). This difference stems from a difference in purpose. The user tests are predominantly aimed at finding the main 80% to 90% of weaknesses in a new design. It has been shown experimentally that this goal can be obtained with five to six observations (Nielsen and Landauer, 1993; Nielsen, 2004a, 2004b) (see also Appendix D). Our intentions are somewhat similar: we want to find out from practitioners if our new method adds value to the definition process of e-services in a multi-channel context. And if so, where it adds value, and where it does not.

6.2 Field experiment details

Our structured field experiment consists of a number of steps, which we describe in this section and summarize in Figure 6-1. In the following sections we explain the various building blocks of this figure.

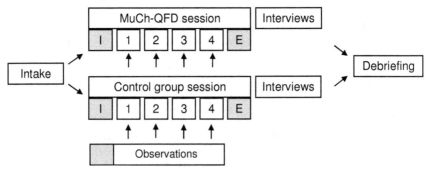

Figure 6-1 Field experiment: I = Introduction, 1-4 = Design activities, E= Evaluation

6.2.1 Intake phase

Each e-service definition case started with an intake, which had three purposes. Firstly, we wanted to find out whether a case would be suitable for our test purposes. This means that there is an e-service idea that is still in the initiation phase as defined in section 4.2, that has a multi-channel context, that is interesting enough for participants (to be motivated to participate) and that still leaves questions unanswered after the intake. Secondly, explaining our procedures and managing the expectations of the 'idea owner' that initiated the e-service design case. And thirdly, during the intake a number of questions had to be answered to allow for well-prepared e-service definition sessions. Also, the different roles required in the sessions were identified, and suitable participants were selected. Our intake protocol was standardized and is discussed appendix C. In short, the intake questions addressed an outline of the e-service idea, why it would be valuable to target customers, added value in relation to existing e-services and offline services, comparisons with competitive initiatives, financial benefits and costs.

6.2.2 E-service definition sessions

As stated before, each case team was split into two groups, with one half attending a MuCh-QFD session, the other half attending an FE session. Each case was initiated by an 'idea owner'. This person not only introduced the case idea to us, but also had an important part in the intake phase and preparation of the sessions, for instance by helping us select the session participants. Each session had at least four participants. The four participants were selected to take a different perspective during the session. As explained in section 4.2, there were three perspectives that always had to be present: customer, channel and supplier focus. The customer focus was represented by sales- or marketing professionals or, in case of an SME, a director who had customer contact on a regular basis. The supplier focus was generally split (and represented by two different people in each session) into a marketing focus and a focus on ICT and processes. The fourth role addressed the perspective of other channels and was represented in consultation with the 'idea owner'. In some cases it was represented by an intermediary (insurance agent), and in other cases by customer service, or by a product manager who would represent the main support tasks via other channels. As explained in chapters 1 and 3, the sessions focused on defining 'auxiliary services' around a core product or service. Although these auxiliary services (like product education, or after sales product support) can be very important for the overall proposition, they are generally not the core proposition that the customer pays for (Grönroos, 2000). Hence, each session took one specific core product or package as a basis for discussion, and defined auxiliary service support around it.

From an experimental viewpoint, similarity and comparability between MuCh-QFD and FE sessions was important. Hence, we made sure that for both session types:
- The sessions took an equal amount of time;
- The same electronic group decision support system (GDSS) is used: GroupSystems™

- The same problem introduction presentation is used (except for explanation of the design method used);
- The agenda formats and types of activities are mostly similar: e.g. individual input versus group discussions versus breakout exercises, number of breaks, duration of agenda items etc;
- And as a specific extension of the previous point: both sessions consisted of four phases;
- The same moderator guides all groups (both session agendas are highly structured and pre-programmed with the aim of ensuring that the results are more the effect of the methods used and less the effect of the moderator or coincidental group effects);
- The moderator removes issues that need further investigation from the session (e.g. regarding customer priorities or technical possibilities) to a task list for later consultation. This allows for progress through the methods and comparability between the sessions;
- As a way to reduce 'group-think' or session dominance by one participant, there are many activities based on individual inputs and individual judgments (in addition to separate group discussions to facilitate group creativity);
- Both sessions (for one e-service case) are conducted on the same day; the length of the break ensures that morning session participants leave before the afternoon participants arrive; which method is used in the morning and which for the afternoon roughly alternates per case (practical issues come into play here, which we discuss below).

Apart from these issues regarding experimental design, there were also several practical issues that had to be resolved:
- Part of our MuCh-QFD method was already present during intake. For example, we specifically asked what the advantages of the new e-service would be for customers, in relation to existing e-services, in relation to competitor offers, in relation to the services offered via other channels. We also explicitly mapped customer processes onto the cycle of pre-sales, sales and after-sales activities, on the basis of which we made choices regarding the e-service elements on which to focus. Hence, to avoid interactions between MuCh-QFD elements from our intake and the control sessions, we aimed to put the participants who had conducted the intake (which were generally also the idea owners) into the MuCh-QFD session. As a trade-off, this introduces a certain disturbing factor: one may argue that our MuCh-QFD measurement results are not the effect of the method, but of the presence of the idea owner. In our results chapter (chapter 7) we will check for this (not all idea owners were in the MuCh-QFD session, which gives us some basis to discriminate between the two effects).
- Due to practical considerations (see also previous point) we were not entirely free to decide which type of session would take place in the morning and which in the afternoon. As a result, we had four cases starting with the MuCh-QFD session, and two starting with the FE session.

6.2.3 Observations during sessions

During each session we employed observers to register the extent to which the design requirements were met during each part of a session. They used the measurement protocol discussed in appendix E. Hence, for each of the four agenda parts of each session, we generated measurements on the degree to which customer orientation, channel coherence etc were discussed. Each observer would be present the whole day and score a morning as well as an afternoon session.

Experience with observers has shown that a certain degree of training increases their scoring reliability (Holsti, 1969). In this respect, we had two groups of student observers. The first, and largest, group of observers knew both sessions, had been pre-test participants and had attended a practice session as observer. An advantage of this group is the degree of experience; a disadvantage is that they may be biased because they know which group is the control group. The second group of observers received a briefing on both sessions (without knowing which group was the control group) and on the observation protocol. Both groups started observations by scoring 'part 0' of the sessions. This was the first session introduction (goals, agenda, e-service description and preliminary results from the intake) which lasted 30 minutes, during which participants were invited to respond to and discuss the expected value of the service, costs and benefits, and competitor activities. We consider the scores on part 0 as a training session for the observers, not as a measurement. An advantage of the second group of observers is that they are totally unbiased regarding both sessions. A disadvantage is that they received less training than group one. In our analysis we found some interesting differences between both observer groups, which we discuss as part of the reliability and validity analysis of our measurement instruments in section 6.4. The singularities that we found were not serious enough to make the observations biased.

6.2.4 Questionnaires and interviews at end of sessions

Participants knew that they were in an experimental setting, but they did not know in advance which methods we would use in the sessions. They were told that we were comparing two different design methods, and that we could not provide any details in advance. This also meant that none of them had any feeling of being in a control group session. At the end of each session, participants were asked to answer fill in the questionnaire. For details, see appendix E. In summary, each questionnaire had the following sections:
- Background questions (previous experiences, expectations etc);
- Subjective judgments regarding usefulness of different sessions elements;
- A set of objective events questions, addressing per requirement whether events had occurred or not according to participants. These questions formed the main measurement instrument we used with the business participants to record performance on our design requirements. For details on these questions (see section 6.3).

Immediately after they completed the questionnaire, the participants were interviewed. The interviews had two purposes. Firstly, to ensure that questions had been answered correctly, and secondly, to receive additional feedback on the sessions. A short report was made of each of these interviews.

6.2.5 Debriefing meeting one to two months after session

Approximately one to two months after each e-service case, the results were discussed with the participants. Because this occurred on a voluntary basis, only 60% of participants were present during the debriefing sessions. A debriefing lasted 1.5 hours: the first hour was used to discuss the results of the two sessions, explain the results to the participants of the other sessions, discuss similarities and differences, and discuss next steps. The last 30 minutes were used to generate feedback on our approach. To guide this feedback, we asked six questions. Each question could be answered with yes, no or 'neutral/maybe/don't know'. The questions were addressed one at a time: first the answers were collected by raising hands, and then the participants were invited to share their opinions with the group.

The six questions were: a) Do you think the resulting e-service definitions are valuable? b) Did the sessions help specify the e-service or its market positioning? c) Do you think the sessions are complete in identifying the main issues for this new e-service, considering the time spent? d) Do you think the intake and/or the service matrix discussions with stakeholders help speed up service innovations? e) Are you going to use the results? f) Would you consider using one of these sessions again in the future?

The group discussions following these questions were useful for providing insights into the overall opinions of the participants after one or two months.

6.3 Development of measurement instruments

Alongside collecting qualitative data, we developed two formal measurement instruments, based on the design requirements identified in chapter 3: 1) a questionnaire that participants filled in immediately after a session, and 2) an observation protocol that external observers used during sessions. These are used next to a) our intake interviews per case, b) the participant interviews immediately following the sessions, c) the responses we collected in our debriefing sessions one to two months after the sessions, and d) next to our own observations per case. To our knowledge no validated scales existed for our measurement purposes, which is why we created our own measurements instruments. A first validation step was using pre-tests (see also section 5.4): 'do participants understand the questions and consider them relevant?' and 'do preliminary quantitative results show differences in scores for the design requirements between both sessions?' As a second validation step we looked at the reliability of items that indicate the core concepts and at cross-measurement reliability (see section 6.4): 'to what extent are the results from the two instruments for our design requirements similar?' An overview of the items per construct and per measurement instrument is given in Table 6-2, and below we discuss the development of each instrument.

6.3.1 Development of participant questionnaires

We used two questionnaires, one for each session type. Most questions were the same. The only differences occurred at the point where participants were asked to express how useful they thought the various agenda items of the sessions were (questions 9 and 10, see appendix E). As explained in 6.2.4, the questionnaires we used had the following sections:

- Background questions (job position, years in the industry, previous experiences, expectations etc);
- Subjective judgments regarding usefulness (of the various session elements, for varying purposes) and check questions to measure the degree of satisfaction and the degree of awareness of problem complexity (which both may influence the other scores, see our pre-test results in section 5.4);
- A set of questions regarding 'objective' events, addressing whether, according to the participants, something had occurred or not.

We briefly discuss the first two sections, and then discuss the questions regarding 'objective' events, which provide us with the main results for our statistical analysis.

First, we asked several questions with regard to job position, years in the industry, previous experiences, expectations and more subjective judgments. They are intended to capture some of the relevant characteristics and expectations of the participants. In addition, some of the questions were used as control variables, to assess disturbing 'satisfaction-' and 'awareness effects' that we observed in our second pre-test round (see also section 5.4.2). Questions 5a to 5c ('meets my expectations', 'in comparison to my other experiences, I am satisfied' and 'meets the desirable level') were taken from a validated customer satisfaction scale (Wang, Po Lo, Chi and Yang, 2004). We used these questions to measure 'satisfaction effects'. Question 8a ('session made me realize service dilemmas') was used to measure 'awareness effects'. Thirdly, other questions on usefulness were designed to measure overall satisfaction, as well as the satisfaction per design requirement and per agenda activity. They were used to check whether there are significant differences between the two session types with regard to these elements.

The third and final section of our questionnaire is the set of questions that measure objective events related to the seven design requirements from 'customer orientation' to 'stakeholder communication'. As explained in chapter 4, requirement 8 - 'coherent concept communication during implementation' - is not included in our research design, because we exclude the implementation phase from our tests. For each design requirement we defined a set of items, designed to register the more objective occurrence of events during the session, as experienced by the participants. For example, regarding customer orientation, the first two items were: 'Did the team discuss what the top 5 customer needs are?' and 'Have customer needs been prioritized?' These items were used as index scales, generating maximum scores of 6 to 9 points for each requirement (for details see appendix E).

The development process for this instrument can be characterized by a first, conceptual phase, and a second, (pre)test phase. Below we describe the pre-test results; here we describe the conceptual phase. The conceptual phase consisted of the following steps: a) First we developed constructs for each design requirement. We formulated a set of items, based on input from design literature and our case experiences. These items were formulated as yes/no questions, and each yes score counted as 1 point. Per construct up to 10 points could be 'earned'. Hence we created preliminary additive scales per design requirement. b) A subsequent step of construct development focused on prioritization and line of reasoning per requirement. c) Next, we moved to the overall perspective: the overall construct set was reviewed for completeness and overall balance in requirements and items. This resulted in our original set of 10 x 8 items (for eight design requirements) displayed in appendix E. These constructs aimed at a very high level of design process performance and were formulated to be method-independent. d) The next step was to review the constructs with regard to usefulness in terms of measuring possible differences between our sessions. In this step several items were removed that were too 'advanced' for our session. For example, after we had chosen to focus the session on defining functionalities, and not include detailed Internet site layout, navigation structures and usability, several items could be dropped (for example, 'customer process designed for customer friendliness?' and 'more customer friendly than competitors?') Also, the items belonging to requirement 8 were removed in this step, since requirement 8 was outside the scope of our research design. In the following section we describe how the resulting measurement set was used and refined via pre-tests.

6.3.2 Pre-testing participant questionnaires

Participant questionnaires were tested in our second and third pre-tests (see section 5.4 for more details on the pre-tests). In the second pre-test, our questionnaire was close to its final form. As a consequence of the second pre-test, several changes had to be made. For most requirements, this meant changing the wording of items. For two requirements, channel synergy and focus, we significantly changed our constructs. Firstly, regarding channel synergy, several items were non-discriminatory. Also, several items needed to be rephrased to create greater focus on multi-channeling and/or design process observations. Secondly, regarding focus, it proved to be difficult to carry out discriminatory measurements. Because the scores were exceptionally high for both methods (7.0 to 7.4 out of 8), we decided to leave out the items that proved to be non-discriminatory. Finally, we decided to introduce control questions (questions 5a to 5c, and 8a) for the 'satisfaction' and 'awareness effects' we found in our second pre-test (see section 5.4). These questions are meant to check whether differences are very pronounced across cases or design conditions as far as satisfaction levels or degree of awareness of problem complexity are concerned, and we use them in chapter 7 to check for interaction effects.

In the third pre-test the new questionnaires were tested, and since we found no new problems we decided to leave them as they were.

6.3.3 Development of observation protocol

We used observers in all sessions to register to what extent the design requirements (customer orientation, channel coherence etc) were met. The same observation protocol was used for both session types. The complete measurement protocol is discussed in appendix E. Here, we explain how we developed our protocol.

The observation protocol was derived from the objective event questions of the participant questionnaire. This was done between the second and third pre-test. In this step, the wording was condensed, all questions were put on a single page, and several sub-items were left out, the main reason being that we wanted to reduce the complexity by reducing the overall number of items (we reduced the number of items from 47 to 37, divided among seven requirements). In this process, we sometimes combined two nuanced items into one simpler item, and some items were simply deleted because we thought they offered the least added value. Regarding our speed requirement, expressed as 'progress in the session', we found that we had to replace the way it was included in the questionnaire. Whereas the new observation items were very much aimed at progress for each element on the agenda ('reach the desired result?', 'result orientation of team?' and 'efficient way of working?') the questionnaire items were aimed more at checking to what extent certain results had been obtained by the end of a session ('customer needs addressed?', 'value in relation to competitors addressed?' etc). Hence, we replaced the eight 'achievement check' items of the questionnaire by the three 'process' items mentioned above.

6.3.4 Pre-testing observation protocol

As a second step in the development of our observation protocol, we tested them in our third pre-test round (with five and six observers per session respectively, see section 5.4.3 for more details). We collected observer feedback, which resulted in some rewording but did not change the items we used.

What did change as a consequence of the pre-test was the level of detail in following the session agendas. In pre-test the observations were performed at a very detailed level: the MuCh-QFD agenda consisted of 18 activities, hence 18 observation sheets had to be scored. Observers complained about this level of detail: they found it cumbersome, and we feared this would reduce measurement reliability. Also, from our experimental point of view we concluded that a higher level of abstraction, clustering and measuring in four main agenda parts per session made more sense. It followed the internal logic of the sessions more accurately. Hence, we decided to use our observation protocol to generate four measurement episodes per session. Thus, each agenda part (from I to IV) coincides with one measurement episode, see Table 4-1 for the MuCh-QFD agenda and Table 5-5 for the FE agenda.

Another complaint was that it was sometimes hard to decide whether an answer was really 'yes' or 'no'. Some observers argued that a 'neutral/partly addressed' category might be suitable. Other observers, however, indicated that this would make the protocol much harder: to keep track of 37 items in parallel, as well as provide a nuanced score, would be too much. From our point of view, we thought it might be good to encourage people to make up their minds, to prevent them from giving lukewarm answers, so we decided to maintain the yes/no dichotomy.

Observers also complained that participants sometimes said sensible things, but 'there were no corresponding boxes to check'. Although we agreed that this was the case, we decided not to change the protocol. This is a direct consequence of our methodological choices to focus our measurement instruments on those elements where we expect our MuCh-QFD session to show advantages, see also section 6.1.3. Hence, we decided to maintain our scope on the seven predefined design requirements.

Table 6-2 Overview of items per measurement construct (questionnaire versus observations)

Questionnaire items	Observation items
1. Customer orientation: Are customer needs and customer value incorporated into the design?	
a. Five main customer needs b. Ranking customer needs c. Customer needs checked with customers d. Customer needs starting point of design e. Design choices related to customer needs f. Checked if e-service has added value for customer	a. Five main customer needs b. Ranking customer needs c. Customer needs linked to design choices d. Scores on customer needs compared to competitor
2. Channel coherence: Do the e-service and existing channel services provide complementary value?	
a1. Discussed existing e-services on web a2. Discussed existing services other channels b. Discussed (dis)advantages of other channels c. Combining channel advantages d. Discussed `handovers' between channels e. Discussed if web reduces time needed across channels? f. Evaluated possible service solutions	a. Discussed existing services other channels b. Discussed existing e-services on web c. Discussed (dis)advantages of other channels d. Discussed if web reduces time needed across channels? e. Discussed `handovers' between channels
3. Channel synergy: Are assets re-used across channels?	
Discussed if new e-service: a. re-uses existing customer relations b. re-uses existing brand awareness c. re-uses existing logistic processes d. re-uses existing information systems e. re-uses information between channels f. automates supplier tasks via self-service g. reduces the necessity to train employees	Discussed if new e-service: a. re-uses existing customer relations b. re-uses existing brand awareness c. re-uses existing logistic processes d. re-uses existing information systems e. re-uses information between channels f. automates supplier tasks via self-service g. reduces the necessity to train employees

4 Competitive positioning: Does the e-service fit the marketing strategy and improve competitive position?	
a. Explicitly placed e-service in overall marketing strategy b1. Discussed e-services of 1 competitor b2. Discussed e-services of 2 competitors c1. Discussed marketing strategy of 1 competitor c2. Discussed marketing strategy of 2 competitors d. Evaluate what (not) to do same as competitors e. Checked if e-service fits brand and communication f. Checked retention or win back of customers g. Checked financial benefits	a. Explicitly discussed e-service in overall marketing strategy b. Discussed e-services of competitors c. Discussed marketing strategy of competitors d. Evaluate what (not) to do same as competitors e. Checked if e-service fits brand and communication f. Checked retention or win back of customers g. Checked financial benefits
5. Progress: How much progress did the team make within the timeframe of the session?	
a. Discuss customer needs b. Discuss value for customer compared to competitors c. Discuss relationship between strategy and customer value add d. Discuss other customer contact points along web site e. Discuss the way channels work together f. Create alternative service solutions g. Discuss fine tuning of service solutions h. Evaluate the technical (im)possibilities	a. Did team achieve desirable result? b. Focused on reaching results? c. Team efficient?
6. Focused design process: Does the team remain focused, despite uncertainties or differences in opinion?	
a. Decisions made together by team b. Clear to everybody which issues to discuss c. Clear to everybody what expected output was d. Structured discussions	a. Decisions made together by team b. Clear to everybody which issues to discuss c. Clear to everybody what expected output was d. Structured discussions
7. Stakeholder communication: Do the various stakeholders understand each other and integrate their perspectives?	
a. Agreed on customer needs as starting point b. Agreed on competitor position as starting point c. Agreed on other channels as starting point d. Agreed on consequences of customer needs e. Agreed on consequences of competitor position f. Agreed on consequences of other channels	a. Every participant's perspective clear? b. Agreed on customer needs as starting point c. Agreed on competitor position as starting point d. Agreed on other channels as starting point e. Agreed on consequences of customer needs f. Agreed on consequences of competitor position g. Agreed on consequences of other channels

6.4   Reliability and validity

In this section we review our measurement instruments. First we perform a reliability check on our observation instrument by checking cross-observer reliability (section 6.4.1). Then we evaluate the effects of using two different groups of observers (section 6.4.2). In section 6.4.3 we evaluate the objective events questions of our participant questionnaire. And in section 6.4.4 we discuss the reliability of both instruments by comparing their results and performing a face validity check in relation to our qualitative case data.

6.4.1   Cross-observer reliability

A well-known problem when using observers is the question of how reliable their scores are (Holsti, 1969). A first test that can be performed is the degree of consensus between observers, which is discussed in this section. We also performed a face value check on the observation scores: are they in line with our expectations? This led to a redefinition of construct 4, on competitive positioning. In 6.4.4 we check consistency of observer scores with questionnaire scores and qualitative case findings per session.

In Table 6-3 we show the degree of cross observer consensus per item. The percentages indicate average consensus between all observers across all four parts of the session agendas. We calculated these averages separately for MuCh-QFD sessions and FE sessions and overall (columns 2, 3 and 4 respectively). When looking at the numbers, one should bear in mind that a 50% score equals no consensus. For an individual session this means that if four people take part, two of them have scored a 'yes' and the other two a 'no' for an item. Hence, a score below 50% is impossible (even if all observers would have made 'the wrong' observation).

The fact that consensus scores are relative scores had an implication for our calculation procedure. It meant that we first had to calculate consensus averages per session, and then average that score across sessions (otherwise 100% 'yes' scores for half the sessions plus 100% 'no' scores for the other half of the sessions would generate a 50% score instead of the factual 100% consensus).

Finally, there was the question as to which items to include in our further analyses. We decided to be conservative: all items that had less then 80% overall cross-observer consensus were discarded. This strictness implies that if we do find significant differences in our analyses, they are not likely to be caused by random effects in the data. A disadvantage of being so strict was that construct 6, focus, became a single-item construct. We decided to keep it for completeness and possibly comparison with our questionnaire results, realizing that findings must be interpreted very carefully. The right column in Table 6-3 shows which items we kept (marked by an 'x') and one column to the left shows gray highlights for consensus scores below 80% of the discarded items.

Table 6-3 Degree of cross-observer consensus for constructs 1 to 7 (n=56)

Requirement Questions	MuCh-QFD consent	FE consent	Overall consent	Included
1. Customer orientation				
a. Five main customer needs	80%	86%	83%	x
b. Ranking customer needs	86%	91%	88%	x
c. Customer needs linked to design choices	73%	82%	78%	
d. Scores on customer needs compared to competitor	91%	97%	94%	x
2. Channel value and coherence				
a. Discussed existing services other channels	85%	84%	84%	x
b. Discussed existing e-services on web	79%	88%	83%	x
c. Discussed (dis)advantages of other channels	86%	90%	88%	x
d. Discussed if web reduces time needed across channels	88%	88%	88%	x
e. Discussed `handovers' between channels	83%	81%	82%	x
3. Channel synergy				
Discussed if new e-service:				
a. re-uses existing customer relations	81%	88%	84%	x
b. re-uses existing brand awareness	90%	91%	91%	x
c. re-uses existing logistic processes	91%	96%	94%	x
d. re-uses existing information systems	84%	82%	83%	x
e. re-uses information between channels	85%	87%	86%	x
f. automates supplier tasks via self service	87%	85%	86%	x
g. reduces the necessity to train employees	99%	100%	100%	x
4a 'Competitor focus'				
b. Discussed e-services of competitors	90%	87%	88%	x
c. Discussed marketing strategy of competitors	88%	92%	90%	x
d. Evaluate what (not) to do same as competitors	88%	89%	88%	x
4b 'Marketing strategy'				
4a. Explicitly discussed e-service in overall marketing strategy	82%	71%	77%	
4e. Checked if e-service fits brand and communication	90%	88%	89%	x
4f. Checked retention or win back of customers	88%	77%	83%	x
4g. Checked financial benefits	90%	89%	90%	x
5. Progression in session				
a. Did team achieve desirable result	88%	88%	88%	x
b. Focused on reaching results?	95%	85%	90%	x
c. Team efficient?	81%	79%	80%	x
6. Focused design process				
a. Decisions made together by team	88%	87%	88%	x
b. Clear to everybody which issues to discuss	81%	68%	75%	
c. Clear to everybody what expected output was	81%	75%	78%	
d. Structured discussions	76%	70%	73%	
7. Stakeholder communication				
a. Every participant's perspective clear?	83%	76%	79%	
b. Agreed on customer needs as starting point	70%	76%	73%	
c. Agreed on competitor position as starting point	88%	82%	85%	x
d. Agreed on other channels as starting point	84%	93%	88%	x
e. Agreed on consequences of customer needs	74%	84%	79%	
f. Agreed on consequences of competitor position	89%	91%	90%	x
g. Agreed on consequences of other channels	88%	95%	91%	x

Overall we can see from Table 6-3 that, with the exception of 6, focus, all constructs remain well-represented with 3 or more items. Only the last two constructs, focus and stakeholder communication, have lost more than one item due to the cross-observer consensus check, and for the latter there are four items left.

We also looked at the average scores of the observers on individual items for MuCh-QFD and FE sessions, to see if they were in line with our expectations. Note that these average scores are not the same as the degrees of consensus displayed in Table 6-3, for example item 3g, reducing necessity to train employees, had virtually a 100% consensus with an average score below 0,10 for MuCh-QFD as well as for FE sessions. An interesting effect occurred with regard to requirement 4, which was competitive positioning. About half of the items (items 4a, 4e, 4f, and 4g) had higher average item scores for FE than for MuCh-QFD sessions. This ran counter to the overall trend across all items, and also to the other items for the construct (4b, 4c, 4d) where on average MuCh-QFD scored higher. A closer look at the wording of the items reveals that the latter items all address 'competitor focus' and the first four items address more general 'marketing strategy' issues. Thus, at face value we decided to split construct 4 into 4a, 'competitor focus' and 4b, 'marketing strategy'. Interestingly, participant scores showed exactly the same split across items (see section 6.4.3), confirming the prudence of our face value analysis via cross-instrument consistency.

6.4.2   Differences between observer groups

As explained in section 6.2.3, we have used two types of observers. The first, and largest, group of observers was familiar with both sessions, had been a pre-test participant and had experience being an observer. An advantage of this group is the degree of experience, which generally improves scoring reliability (Holsti, 1969). A disadvantage is that they may be biased, because they know which group is the control group. The second group of observers received a briefing on both sessions (without knowing which group was the control group) and on the observation protocol, but had no practical experience. The only practice they had was during the introduction of the session, before part I started. We asked them to score that activity too, without telling them that it would not be used in our analyses. An advantage of the second group of observers is that they are totally unbiased regarding the two sessions. A disadvantage is that they received less training than group one. Here we want to check if singularities have occurred as a result of differences between the two observer groups.

For each of our constructs, we took the sum across all four agenda parts per observer and checked on the basis of the Mann-Whitney test if there was a significant difference between the two groups. The 'prior experience' group shows lower scores on requirements 1 to 4b (for FE and MuChQFD) and higher scores on requirements 5 to 7 than the group without prior experience. It is interesting to notice this overall trend, although not all differences are significant. The differences are significant for customer orientation (Mann-Whitney U=231; Z=-1.98; p<0.05), channel synergy (Mann-Whitney U=230; Z=-1.99; p<0.05) and focus (Mann-Whitney U=148, Z=-3.80; p<0.001). The lower scores of the inexperienced group on process requirements may be caused by their lack of experience: especially during breakouts at which much discussion took place they scored relatively low, which may be caused by their unfamiliarity with the sessions, causing them difficulties in perceiving progress, focus and communication. It may also be that the prior experience group is biased: they know that a) the outcome-related requirements 1 to 4b are only expected to occur on selective occasions and may be reluctant to score them on other occasions. And they know that b) the process related requirements are expected to occur in all agenda parts. As a result, they may have been 'too optimistic' in scoring those. Our design is not suited to determine which of these assumptions is valid.

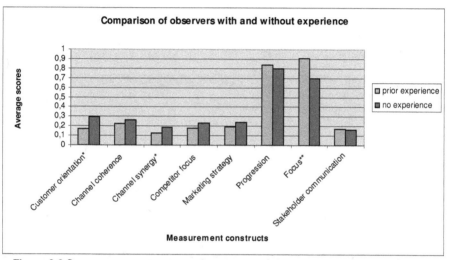

Figure 6-2 Scores per construct per observer group (n=56; ** = p<0.01; * = p<0.05)

To check how much these differences may have influenced our measurements, we performed analyses of variance for each of the design requirements with regard to prior experience versus no prior experience and MuCh-QFD versus FE. None of the analyses showed any interaction effects. The linear effects we found were in line with our previous findings based on Mann-Whitney tests. We also checked the individual scoring per observer to see if there were any differences in terms of the peak patterns in terms of the various agenda parts, which was not the case. Hence, we conclude that, with regard to our main experimental factor, design method (MuCh-QFD versus FE), the measurement disturbances caused by (lack of) previous experience are probably limited. Since observations were paired (each observer scoring an FE and a MuCh-QFD session), differences in observer experience work in similar directions for FE and MuCh-QFD sessions. Based on this we decided to include all the observers in our subsequent analyses.

6.4.3 Participant questionnaire

Table 6-4 MuCh-QFD (n=24) and FE (n=25) scores, stand. dev. and significance

Requirement Questions	MuCh-QFD Avg.	Std.	FE Avg.	Std.	Sig.	MW	Sig.
1. Customer orientation							
a. Five main customer needs	0,96	0,20	0,56	0,51	**	181	0,00
b. Ranking customer needs	0,96	0,20	0,60	0,50	**	193	0,00
c. Customer needs checked with customers	0,17	0,38	0,16	0,37		298	0,95
d. Customer needs starting point of design	0,92	0,28	0,52	0,51	**	181	0,00
e. Design choices related to customer needs	0,88	0,34	0,44	0,51	**	170	0,00
f. Checked if e-service has added value for customer	0,88	0,34	0,52	0,51	**	194	0,01
Construct score	**0,79**	**0,13**	**0,47**	**0,25**	**0,00**	**77**	**0,00**
2. Channel value and coherence							
a1. Discussed existing e-services on web	0,88	0,34	0,52	0,51	**	194	0,01
a2. Discussed existing services other channels	0,75	0,44	0,44	0,51	*	207	0,03
b. Discussed (dis)advantages of other channels	0,71	0,46	0,36	0,49	*	196	0,02
c. Combining channel advantages	0,58	0,50	0,28	0,46	*	209	0,03
d. Discussed `handovers' between channels	0,42	0,50	0,32	0,48		271	0,49
e. Discussed if web reduces time needed across channels?	0,54	0,51	0,56	0,51		295	0,90
f. Evaluated possible service solutions	0,50	0,51	0,48	0,51		294	0,89
Construct score	**0,63**	**0,28**	**0,42**	**0,28**	**0,01**	**177**	**0,01**
3. Channel synergy							
Discussed if new e-service:							
a. re-uses existing customer relations	0,83	0,38	0,80	0,41		290	0,77
b. re-uses existing brand awareness	0,42	0,50	0,52	0,51		269	0,47
c. re-uses existing logistic processes	0,42	0,50	0,48	0,51		281	0,66
d. re-uses existing information systems	0,79	0,41	0,68	0,48		267	0,38
e. re-uses information between channels	0,75	0,44	0,52	0,51		231	0,10
f. automates supplier tasks via self service	0,63	0,49	0,64	0,49		296	0,91
g. reduces the necessity to train employees	0,13	0,34	0,08	0,28		287	0,61
Construct score	**0,57**	**0,28**	**0,53**	**0,27**	**0,63**	**277**	**0,63**
4a 'Competitor focus'							
Discussed what competitors do with e-services:							
b1. 1 competitor	0,63	0,49	0,40	0,50		233	0,12
b2. 2 competitors	0,36	0,49	0,32	0,48		263	0,76
Discussed competitor market strategy							
c1. 1 competitor	0,33	0,48	0,16	0,37		248	0,16
c2. 2 competitors	0,22	0,42	0,12	0,33		260	0,37
d. Evaluate what (not) to do same as competitors	0,21	0,41	0,12	0,33		274	0,41
Construct score	**0,34**	**0,34**	**0,22**	**0,26**	**0,23**	**243**	**0,23**

Requirement Questions	MuCh-QFD		FE		Sig.	MW	Sig.
	Avg.	Std.	Avg.	Std.			
4b 'Marketing strategy'							
4a. Explicitly placed e-service in overall marketing strategy	0,38	0,49	0,40	0,50		293	0,86
4e. Checked if e-service fits brand and communication	0,33	0,48	0,48	0,51		256	0,30
4f. Checked retention or win back of customers	0,33	0,48	0,56	0,51		232	0,11
4g. Checked financial benefits for supplier	0,63	0,49	0,92	0,28	**	212	0,01
Construct score	**0,42**	**0,30**	**0,59**	**0,30**	**0,03**	**194**	**0,03**
5. Progression in session							
Which aspects are discussed:							
a. customer needs	1,00	0,00	0,92	0,28		276	0,16
b. value for customer compared to competitors	0,83	0,38	0,68	0,48		254	0,22
c. relation between strategy and added customer value	0,54	0,51	0,60	0,50		283	0,68
d. other customer contact points along website?	0,88	0,34	0,76	0,44		266	0,30
e. the way channels work together	0,71	0,46	0,40	0,50	*	208	0,03
f. create alternative service solutions	0,42	0,50	0,32	0,48		271	0,49
g. fine tuning the service solutions	0,33	0,48	0,48	0,51		256	0,30
h. evaluate the technical (im)possibilities	0,17	0,38	0,36	0,49		242	0,13
Construct score	**0,61**	**0,21**	**0,57**	**0,23**	**0,53**	**269**	**0,53**
6. Focussed design process							
a. Decisions made together by team	1,00	0,00	0,80	0,41	*	240	0,02
b. Clear to everybody which issues to discuss	0,88	0,34	0,88	0,33		299	0,96
c. Clear to everybody what expected output was	0,79	0,41	0,72	0,46		279	0,56
d. Stuctured discussions	0,96	0,20	0,88	0,33		277	0,32
Construct score	**0,91**	**0,14**	**0,82**	**0,28**	**0,45**	**268**	**0,45**
7. Stakeholder communication							
a. Agreed on customer needs as starting point	0,92	0,28	0,76	0,44		253	0,14
b. Agreed on consequences of customer needs	0,96	0,20	0,76	0,44	*	241	0,05
c. Agreed on competitor position as starting point	0,79	0,41	0,60	0,50		243	0,15
d. Agreed on consequences of competitor position	0,75	0,44	0,52	0,51		231	0,10
e. Agreed on other channels as starting point	0,83	0,39	0,28	0,46	**	131	0,00
f. Agreed on consequences of other channels	0,83	0,38	0,24	0,44	**	122	0,00
Construct score	**0,84**	**0,21**	**0,53**	**0,33**	**0,00**	**136**	**0,00**

Legend: ** = p<0.01; * = p<0.05

The options for validating our questionnaire (section 6.3.1) are limited. Since our items are not at the right measurement level (yes/no dichotomies are not suitable for factor analysis), factor analysis is not possible. In this section we look at significance levels of differences between MuCh-QFD and FE per item and per construct. We also conduct a face value analysis of the average scores and standard deviations for MuCh-QFD versus FE. In section 6.4.4 we perform a reliability check by comparing questionnaire with observer results and with our qualitative case findings per session, to see if the data are consistent.

Table 6-4 shows from left to right the items that were used, the MuCh-QFD average scores and standard deviations, the FE average scores and standard deviations, whether there was a significant difference, what the Mann-Whitney scores were, and the significance levels. Several things can be observed from the table. Firstly, at item level, 15 out of 47 items show significant differences at the 0.05 or 0.01 level. Secondly, most of these significant differences can be found in constructs 1, customer orientation, 2, channel coherence, and 7, stakeholder communication, which are also the only 3 constructs that show significant differences at the 0.01 level. Thirdly, construct 4, competitive positioning, can be split in the same way as the observer instrument in the previous section. Again, all 'competitor focus' items score higher for MuCh-QFD and all 'marketing strategy' items score higher for FE sessions. The difference between both session types is significant ($p < 0.05$) for the marketing strategy construct as a whole. Fourthly, apart from 'marketing strategy' the general trend is that all constructs score higher for MuCh-QFD than for FE, which is in line with our expectations. Fifthly, there are a number of items that do not follow this overall pattern, each construct contains one or more of these items: they are items that either show no differences between both session types, or are even slightly lower for MuCh-QFD than for FE. Two of those differences are almost significant: 5.g, fine tuning service solutions, and 5.h, evaluate technical (im)possibilities. From our qualitative case analyses we recognize that these issues are hardly addressed at all during MuCh-QFD sessions and more so during FE sessions. The fact that the item output and the qualitative case findings are in accordance in this respect can be seen as a form of confirmation of the validity of our measurement instrument.

Looking back at these five observations, on the one hand we conclude that there is a match between our expectations and case findings on the one hand and the measurement output on the other hand. On the other hand a drawback is that this conclusion is not based on a reliable statistical procedure to remove unreliable items or validate construct composition. In that sense, these constructs remain relatively close to 'raw data'. In the next section we continue to check consistency with observer results and with our qualitative case information per session.

6.4.4 Cross-measurement consistency

In this section we check cross-measurement consistency. Firstly, we discuss quantitative comparison between both formal measurement instruments. Secondly, we carry out a qualitative evaluation of the instrument scores per session, comparing them to each other and our own observations.

It would have been convenient for our cross-measurement consistency check if they could easily be recalculated in relation to each other, like changing a scale from centimeters to meters. However, this is not the case. The information in our two measurement instruments is of a somewhat different nature. Observers measure four peak occurrences (generating, for example, scores of 75%, 50%, 25% and 50% for a design requirement construct) and participants provide overall scores regarding what has occurred (generating for example a score of 60% for the same requirement). Below we discuss what makes the measurements hard to compare and we adopt two quantitative approaches to transform the session-based observations into overall scores. Then we discuss limitations of each quantitative approach and check to what extent observation and questionnaire scores correlate.

To transform observation scores into overall scores we can either take the average across all agenda parts or take the maximum score of the four agenda parts. The first approach has the serious disadvantage that if, for example, channel coherence is addressed fully in one agenda part, but not at all in the others, the overall score will be low, which is not in accordance with the facts. To check this assumption, we added the number, which indeed generated nonsense data. The second approach also has its disadvantages. If we again take the example of channel coherence, and observers generate a 50% score in three of the four agenda parts, what does that mean? Is the overall set of discussions worth only 50% (meaning each discussion has added nothing to the others) or 75%, or 100%? For the sake of completeness we performed these calculations and compared these peak values with participant scores and found two significant Pearson correlations: for requirement 1, customer orientation ($p<0.01$), and 7, stakeholder communication ($p<0.05$). None of the other constructs generated significant correlations, not even the other two constructs that appeared relatively robust in the previous section: channel coherence and marketing strategy. We attribute the fact that few significant correlations are found to the limited number of observations (only 12 sessions to compare scores) and the inherent differences mentioned above between the peaks we observed and the overall score, which makes comparison difficult. Overall, we can conclude that constructs 1, customer orientation, and 7, stakeholder communication are relatively robust.

The second approach to checking instrument consistency is a face value triangulation of a) observer scores, b) participant scores and c) our qualitative case analyses per session. In Appendix G we have provided details per case and session. The basic question we address is: Do the scores per session have face value validity in relation to each other and to the case analysis information? If the answer is yes, the measurements are relatively robust.

The answer to this question can of course vary per construct. In Table 6-5 a summary is presented of the evaluation of the session-based instrument scores in relation to each other and to the qualitative case information per session. In the right-hand column we provide our assessment regarding the reliability of the constructs. This column shows that questionnaire scores awarded by the participants deviate from our case analysis findings more than the observer scores. With regard to this difference, we perceived a mild underlying structural effect. Observers stay closer to 'the letter of the law' by factually scoring whether

117

or not something has been discussed. Participants appear to stay closer to 'the spirit of the law' by scoring items only if they have the feeling that the items have been addressed sufficiently as a team. This is most visible with constructs 4a, competitor focus, and 5, progress, for cases B, insurer portal, D, offer registration, and F, telecom bundle, in MuCh-QFD sessions, and in general in case C, intermediary portal, for the FE session where (valuable) dominance of one person resulted in a lower participation by others and in lower questionnaire scores.

Table 6-5 Instrument consistency analysis via face value triangulation of questionnaires, observer scores and qualitative case information per session

Requirement	Observer and participant scores across sessions	Instrument Consistency
1. Customer Orientation	All scores are in line with each other and case information per session. Only case C, intermediary portal, session FE was more customer-oriented than observer and questionnaire scores express[16].	**High** (Obs.: 12 of 12) (Quest.: 12 of 12)
2. Channel Coherence	Most scores were in line with case information per session. Except three of the questionnaire scores: for case B, insurer portal, MuCh-QFD scored lower and FE scored higher than our case analyses, and FE for case E, online billing, scored lower.	**Obs. High** (Obs.: 12 of 12) **Quest. Moderate** (Quest.: 9 of 12)
3. Channel Synergy	Most scores were in line with case information per session. Except two of the questionnaire scores: MuCh-QFD for case A, absence management, scored higher than our case analyses, and FE for case B, insurer portal, scored higher.	**Obs. High** (Obs.: 12 of 12) **Quest. Moderate** (Quest.: 10 of 12)
4a Competitor Focus	Most scores were in line with case information per session. Except three of the questionnaire scores: MuCh-QFD for cases B, insurer portal, D, offer registration and F, telecom bundle, scored lower than our case analyses, and lower than observer scores. Participants structurally appear to recollect fewer competitor considerations than external observers score.	**Obs. High** (Obs.: 12 of 12) **Quest. Moderate** (Quest.: 9 of 12)
4b Marketing Strategy	All scores are in line with each other and case information per session. Only case C session FE shows relatively low questionnaire scores[17].	**High** (Obs.: 12 of 12) (Quest.: 12 of 12)
5. Progression	For this construct, the questionnaire and observer items are totally different. Hence, the nature of comparison is different from the other constructs. The questionnaire construct appears to capture progress less consistently: MuCh-QFD for cases B, insurer portal, and D, offer registration, scored lower than our case analyses, and than observer scores.	**Obs. High** (Obs.: 12 of 12) **Quest. Moderate** (Quest.: 10 of 12)

---

[16] However, this is not caused by measurement unreliability. The scores did express correctly what happened in the session, but our items are defined quite narrowly. This means that scores depend on explicitly incorporating customer needs as integral part of the design process. If there is no customer-oriented group process, but e-service definitions are customer-oriented due to one dominant participant, as was the case in this session, our construct expresses the first and not the second effect.

[17] Again, this is not caused by measurement unreliability. The group process on the subject was limited in this session, but marketing strategy issues were addressed.

Requirement	Observer and participant scores across sessions	Instrument Consistency
6. Focussed Design	All scores are in line with each other and case information per session. However, the observer construct consists of only 1 item. This limits robustness and increases variance. Despite these limitations it appears relatively robust[18].	**Obs. Moderate** (Obs.: 12 of 12, but single item) **Quest. High** (Quest.: 12 of 12)
7. Stakeholder Communication	Most scores were in line with case information per session. Except for the observer scores of case B, insurer portal, for MuCh-QFD, which were lower than participants and our case analyses indicated.	**High** (Obs.: 11 of 12) (Quest.: 12 of 12)

Legend: High = 11 or 12 sessions consistent; Moderate = 10 or 9 sessions consistent.
Obs. = Observer instrument; Quest. = Questionnaire instrument

Table 6-5 also shows that most measurements are, with some exceptions, at face value similar to the case information per session and mostly in line with each other. The right-hand column displays our final reliability assessment for each construct.

6.5 Conclusion

In this chapter we discussed how to evaluate the performance of our new design support method on the design support requirements. First of all, we decided to use an experimental design similar to 'static group comparisons' (Hagenaars and Segers, 1980). This means that we use the form often used in medical tests, with one half of our participants exposed to the experimental condition - MuCh-QFD sessions - and the other half to a control condition - FE sessions. To ensure a maximum degree of comparability between the two groups, we randomly split each design team into two groups: half of them attended the MuCh-QFD session, the other half attended the control group FE session. Both were conducted on the same day and participants from both groups were not allowed to contact each other in between the sessions, to avoid influencing the afternoon sessions. We developed two formal measurement instruments that used constructs that matched each of our design requirements (customer orientation to stakeholder communication): firstly observation protocols that generated four measurements during the four agenda parts that each session had, filled in by observers, and secondly questionnaires that participants filled in immediately after the sessions. In addition to these formal measurements we also used other observations, which enabled us to conduct qualitative case analyses. We collected case information during the intake that precedes the sessions. We logged design process observations during the four agenda parts of all sessions. We interviewed all the participants individually, immediately after they completed their questionnaires, to obtain additional information and insights. In addition to the formal constructs, in the questionnaires the participants were also asked to provide evaluations: opinions about the sessions, their satisfaction, previous experiences and expectations they had prior to the session etc. Finally, we conducted debriefings

---

[18] As a check for robustness, we compared the single-items scores with construct scores if all items were included. Although this lowered scores somewhat, it did not change the overall patterns. Hence, there is some robustness between the single item construct and the overall construct.

between one and two months after each session, during which we not only presented the results, but also collected feedback on the sessions.

These different data sources have allowed us to perform triangulation, to check the consistency across our measurements. Besides, we also checked reliability per instrument after collecting the data. It is interesting to note that our fourth construct - for 'competitive positioning' - had to be split into 'competitor focus' and 'marketing positioning', and that the data from the two formal instruments suggested the same split. Besides, the split was confirmed via a face value comparison with our qualitative case material.

7 Results

If you torture the data enough, they will confess.
(Ronald H. Coase, Nobel Prize Laureate in Economic Science)

Case Exhibit 7-1 Case illustration of design method impact

Case F: telecom bundle, had an e-service idea that was still in its early stages. The intake really helped form the ideas about what the e-service should and should not be. The intake also helped create the necessary focus for the session. Part of the challenge was the breadth and complexity of the e-service, which encompassed the promotion and selling of three quite different core products in one package. In the FE session the group encountered particular difficulties in defining the core solutions on the website. The free format approach of FE session part III, in combination with the fact that nobody as yet had strong ideas about the e-service, made progress difficult. In the MuCh-QFD session in particular the value of 'multi-channel' part III became apparent, and participants were surprised by the multi-channel aspects that played a role in service success. They were glad this was part of the agenda. By contrast, multi-channel issues became less apparent in the FE session of case F.

In this chapter we answer our sixth research question:
6. How do the e-service definition methods perform with regard to the design support requirements?

To answer this question we start with a qualitative analysis in section 7.1. We subsequently provide case descriptions, perform a cross-case analysis of design processes during intake and sessions, evaluate session outputs, analyze participant opinions and we analyze feedback from the debriefing sessions we conducted some weeks after our design sessions. Section 7.2 is more quantitative, presenting statistical results based on our measurement instruments: participant questionnaires (7.2.2) and observer scores (7.2.3). In this section we also address the results in relation to our hypotheses. Finally, in section 7.3 we address overall findings and conclusions regarding our experimental results. In chapter 8 we present the overall conclusion of our research.

7.1 Qualitative results

In this section we move from observations of the author towards participant observations and opinions. We provide case descriptions in section 7.1.1, we perform a cross-case analysis of design processes in section 7.1.2, we illustrate session outputs in section 7.1.3, and we evaluate outputs in relation to our design requirements in sections 7.1.4. In 7.1.5 we analyze participant opinions and in section 7.1.6 we analyze feedback from the debriefing sessions we conducted some weeks after our design sessions.

7.1.1 Case descriptions

This section provides case descriptions and discusses specific features of each of the 12 design sessions. From an experimental perspective, session-specific singularities may be seen as disturbing the uniformity of conditions of our field experiment. However, from a qualitative perspective these singularities can provide useful insights. The main purpose of this and the following section is to address these singularities and insights.

To provide background information we have summarized the service concepts for each case in Table 7-1. More extensive summaries can be found in Appendix F.

Table 7-1 Description of service to be designed per case

	Basic description of service
Case A Management of employee absence	This service was initiated by a medium-sized firm that solves employee absence problems for other firms. The core product is absence management (due to by illness, accidents etc). The e-service is the online monitoring and (self) management of employee absence. This is a complicated process because of the many legal and financial obligations facing employers. Online transparency and manageability should be improved for customers, which were SME's with 5 to 50 employees. Focus is on settlement and after-care service.
Case B Insurer portal	This service was initiated by a large insurer that sells via intermediaries. The e-service enables customers to buy and modify low-risk insurance policy bundles via the intermediary's website. Although the content of the site is provided and maintained by the insurer, it has the look and feel of the insurance agent. The service is targeted at insurer's customers that buy a bundle of insurance services, like home, car and liability insurance. Customer preference for ease of use and reliability.
Case C Intermediary portal	This service was initiated by a small intermediary firm, seeking cooperation with other intermediaries and insurers to start an intermediary portal. The e-service design focused on offering damage insurances via a portal that is managed by the intermediary sector. It offers insurance policies via a central 'back office' portal and has the look and feel of the intermediary's site. The service is targeted at consumers who select their damage insurance mainly on price.

Case D Offer registration	This service was initiated by a technology service provider in the insurance sector. The e-service enables digital registration of the advice and offering process of high-risk policies, in such a way that the registration trail cannot be modified later on. The focus in on pre-sales: how are the quotation process and registration combined? The service is targeted at consumers looking for life insurance policies.
Case E Online billing for SME's	This service was initiated by a large telecommunication provider. The e-service offers online availability of mobile phone bill data for SME's, with multiple possibilities for analysis. Increasing service for customers is a central objective of the e-service. Target group: SME's with up to 50 employees that use cell phones for voice and data.
Case F Telecom bundle	This service was initiated by a large telecommunication provider. This service should make it possible to choose and buy triple play (TV, Internet access and telephony) products and features online. The design session focus is on supporting orientation and order entry. The e-service targets consumers who choose products based on price and ease of use, and look for service bundles.

Case A: absence management, was interesting in the sense that the 'core product' around which the e-services would be positioned had yet to be completed and was rather complex. By contrast, all other cases involved existing core products that were relatively mainstream. This had a strong impact on the intake phase, which took about three times longer than average. The main challenge was creating focus and determining what to include in the session and what not. In terms of session output a consequence was that in the multi-channel tasks (part III) in the MuCh-QFD session suggestions were made on how to improve the core product and there were discussions on how tasks should be divided between parties involved. This illustrates that part III of MuCh-QFD helps in making the various parties more aware of customer-related issues. In the FE session, defining the service slogan was relatively useful in comparison to other cases: in light of to the complexity of the service it helped to formulate the core intention of the service explicitly.

Case B: insurer portal, involved a large company that had a culture that was strongly adverse to the idea of competing on price. As a result, it was reluctant to provide its customers with price-related information. It was only due to the overall cohesion in the service matrix between customer needs (including a need for price information) and competitor positioning (showing that competitors were much better at providing price information) that this topic came up in the MuCh-QFD session (not in the FE session). Subsequently it was admitted to be an important competitiveness issue, also due to the presence of an experienced sales person representing the customer perspective. This illustrates the fact that a good customer advocate is needed in service definition sessions. In the FE session the discussion on investment readiness (in part III) was particularly useful to identify some of the bottlenecks in the success of an e-service. Finally, in the MuCh-QFD session a manager was present who became a very staunch advocate of the service matrix (as also happened in corporate case E). In both cases the managers said they considered it a useful way of structuring and managing service innovations.

In Case C: intermediary portal, one of the participants of the FE session could be described as 'a traveler from the future'. His company had started planning a similar e-service four years ago, and it had been commercially operational for two years. He provided a great deal of insight into a) what is a clever way to tackle problem XYZ?' and b) what do customers want, buy, and respond best to?' This was rather unique and did not occur in any of the other sessions. His input had a marked impact on the session. For one thing, he strongly advocated customer focus and emphasized the need to connect all service choices to customer priorities. Thanks to this input, output of the FE session was quite customer-oriented, and comparable to the MuCh-QFD results. However, although the first two agenda parts were relatively democratic (caused by the GroupSystems™ approach), from part III on (a more free format breakout activity) his influence became very strong. His input became almost a teaching session: this is what should come first and this what should come second on the site, this is how customers respond to these service functions, this is how an insurance agent can educate customers on self-service, this is how you can outperform the competition etc. His experience far exceeded that of the others, which led to a website solution and competitive position that very much resembled his own e-service. And although certain participants had been skeptic about the added value of part III, the 'teaching session' made it valuable. Overall, the output of the FE session was of high quality, but in terms of process the group effort turned into a 'solo' in the second half of the session. The MuCh-QFD session was of a high quality and it was concluded by a relatively well-rounded discussion on how the e-service should be positioned. However, in our view the output of the MuCh-QFD session was not much better than that of the FE session, and on some points it was even less specific, due to the 'traveler from the future' in the FE session. This pattern is the opposite from with we saw in the other cases.

Case D: offer registration, was interesting in the sense that at the start of the intake it was not customer-oriented. Its aim was to offer some technical functions to insurance agents, without taking into account how the e-service would fit in the commercial offer to the end-customer. Fortunately, the intake questions did what they were supposed to do and made the e-service focus more customer oriented. This supports our opinion that the intake is part of our service definition method, and that asking the right questions at the start of a service definition process leads to a more customer-oriented design. In case D, the participants of the two sessions had very different ideas about how the e-service should be positioned and what the chances were of commercial success. (In the other cases session the outcomes were much more similar). The e-service had to compete with a paper-based solution. In the MuCh-QFD session the paper-based solution scored seriously lower, whereas in the FE session it was stated that the paper-based solution would be the most attractive one for 80% of the intermediaries. This difference was partly caused by different views on the future, and partly by the fact that the participants of the MuCh-QFD looked at customer priorities, while the FE session mostly considered the priorities of the insurance agent. This illustrates a potential strength of the QFD method: customer orientation (which also became apparent in the intake). It also indicates a potential weakness: for this e-service, success also depends on advantages for and adoption by intermediaries. In our tested format, the MuCh-QFD session is less sensitive to these aspects. It would be interesting to see if the method could be adapted to include these aspects better.

Case E: online billing, was based on an e-service that was inherently very customer-focused: the primary goal was increasing loyalty. Also, the first release was already available and the sessions focused on the second release. Hence, the participants had more experience with the e-service than in the other cases. Both elements became apparent during the sessions. Firstly, due to the loyalty objectives both sessions were quite customer-oriented (so in that respect there was little difference between MuCh-QFD and FE sessions). Secondly, the manager present in the MuCh-QFD session was not only positive about the service matrix (like the manager in case B), but he went one step further, most likely due to previous experience with this e-service. He stated that if the company had used this matrix in the first e-service release, his investment priorities on Internet functions would have been different due to the insights the matrix provides in terms of customer priority. Also, he said that this approach should be used as the basis for the entire innovation process management (which is very close to the QFD philosophy). Two other remarks on this case are that 1) at the end of the FE session the participants concluded that the resulting e-service had become too complex for customers and it would not be easy to simplify the e-service, and 2) that the participants of the MuCh-QFD session were surprised by the amount of multi-channel coordination that would be required to make this service successful (which made them appreciate the added value of part III of the MuCh-QFD agenda).

Case F: telecom bundle, an e-service idea that was still in its early stages. The intake really helped make the ideas about what the e-service should and should not be clearer. The intake also helped create focus in the session. Part of the challenge was the breadth and complexity of the e-service, which included the promotion and selling of three quite different core products in a single package. In the FE session the group found it particular difficult to define the core solutions of the website. The free format approach of FE session part III, in combination with the fact that nobody had strong ideas about the e-service yet, made progress difficult. In the MuCh-QFD session it was in particular the value of part III that became apparent, and the participants were surprised by the multi-channel aspects that played a role in service success and were glad these aspects were on the agenda. In the FE session this was less explicit. One final point with regard to this case (which was apparent in both sessions) is the fact that the firm is very reluctant to promote the core proposition on price or even on a (temporary) financial advantage for the customer. Although the participants very much liked the slogan suggested by of the competitors, 'combineer en profiteer' (combine & benefit), they were very reluctant to use anything similar. They said this would not fit their brand, and even though at a factual and strategic level there were benefits in price based promotion, the company's corporate culture seemed to 'prohibit' communicating anything of the kind. We thought that was striking, especially for the MuCh-QFD session, since financial benefit was an explicit customer desire.

7.1.2 Cross-case analysis of service definition processes of intake and sessions

A first step in this section is to compare the insurance cases with the telecommunication cases. Then we evaluate the added value of the intake, the MuCh-QFD session and the FE session in relation to our requirements.

When comparing the insurance cases with the telecommunication cases we have to conclude that in the insurance cases the focus was of a more inherently multi-channel nature: how to support the face-to-face strengths of intermediaries with additional e-servicing? In the telecom operator cases the initial focus was more on an e-service perspective, while multi-channel aspects were initially considered 'something extra' or 'something internal'. This led to bigger differences between MuCh-QFD and FE, with regard to multi-channel coherence. In the telecommunication cases MuCh-QFD helped trigger multi-channel issues, which were then identified to be quite relevant, much to the surprise of the participants.

There were also some differences between SME's and large companies. Firstly, SME's are generally less prone to value formal management tools or methods for new service development than corporate firms: they consider them a necessary evil at best. Secondly, SME's are also more prone to use their own opinions and experiences as a compass for new service development. Corporate participants generally have much fewer contacts with customers in their daily tasks and they rely much more on market research. As a result, large companies are more used to working with formal service definition methods than SME's, they are more likely to benefit from MuCh-QFD than SME's.

We start the evaluation of our design support by looking at the effects of the intake protocol (see also 6.2.1 and Appendix B for details on the protocol). The protocol starts by defining the auxiliary e-service idea, the core product that it is meant to enhance and the customer target segment. Then the main service process steps are defined from a customer's point of view (following service blueprinting basics, but without explicit 'line of visibility' exercises; question 3 of the protocol). These two steps create significant customer orientation and help specify the e-service, as we most notably saw in cases A: absence management, D: offer registration and F: telecom bundle. This adds value especially when the e-service idea is complex and/or still in the early phases of conceptualization. The next step is to determine the relevance of the e-service in relation to market demand and to determine where the focus of the session should lie. This takes place in questions 4 and 5 of the protocol, which determine in what phase of the buying cycle the e-services offer the highest added customer-related and competitive value. That is also the phase on which the design session will focus. This step helps further define the e-service and determine how it should be positioned in the market (cases A: absence management, C: intermediary portal, D: offer registration and F: telecom bundle). With regard to all cases, it helps create focus for the sessions, by stating what is inside and outside the scope of the design problem to be solved. The next step of the intake (questions 6 to 10) discusses multi-channel issues and competitor (e-)services. This step helps in preparing parts III and IV of the MuCh-QFD session: what multi-channel issues can be expected and what are competitors doing? We also use these questions to check whether there are sufficient multi-channel issues to merit the use of MuCh-

QFD. The information concerning competitors is presented to participants during the introductory presentation of the sessions to ensure everybody has the same background information. Presenting examples of what competitor's had done helped make the e-service idea clear to participants in all cases, except case D: offer registration, where there were hardly any competitors worth mentioned. In case D the participants responded by referring competitors that offer something similar to parts of the e-service. In general, presenting session participants with information about competitors encouraged them to ask questions and it increased involvement. Questions 11 and 12 create a high-level business case (costs and benefits). With hindsight we can state that this step was the least important one for the design tasks in the sessions. It does help make the e-service explicit by indicating where the revenues are expected to come from (e.g. sell a very new and low cost product for case A: absence management, defending sales of low interest product for cases B: insurer portal and C: intermediary portal, providing the cheapest way to comply to law for case D: offer registration, reducing churn for case E: online billing, and cross-selling across products for case F: telecom bundle). Finally, it had to be decided who would be the participants of the sessions. Although this is very much a practical matter, it helps to clarify who the main stakeholders are and involve them at an early stage in the service definition process.

Hence, in relation to the requirements the intake helps create customer orientation, it addresses channel issues (although not so much channel synergy), it gathers data needed for competitive positioning, and it creates focus. Some progress in e-service definition is made during the intake, especially when the e-service idea is still in its early stages. Communication between stakeholder perspectives does not actually take place during the intake, although preparations are made by selecting participants that represent the main stakeholder perspectives. As a final remark concerning the intake it is interesting to note that one of the cases that was canceled (see section 6.1.4), benefited so much from the intake that many of the service definition questions had been resolved by the end of the intake. This was also due to the relative simplicity of the e-service idea.

In MuCh-QFD sessions, customer orientation is promoted by using customer priorities as a reference point throughout of the session. Although this proved helpful in all cases, it was especially helpful in cases B: insurer portal, D: offer registration and F: telecom bundle. Channel coherence was considerably stimulated by part III of the agenda, most notably in cases A: absence management, E: online billing and F: telecom bundle. Interestingly enough, channel synergy was less noticeably promoted by MuCh-QFD. Although several inter-channel win-win services were identified in all MuCh-QFD sessions, it turned out that decisions about cost savings across channels by reusing assets were somewhat premature at this stage of the service definition, and that more details would be needed. Discussions regarding competitive positioning focused strongly on comparisons with competing e-services, due to part IV of the agenda. Broader marketing strategy discussions were addressed to a lesser extent. Only cases C: intermediary portal, and partly B: insurer portal, had serious discussions about how to make the e-service successful (e.g. together with other players) and how to position the e-service. This was due to the nature of the ideas in those particular, which were meant to counter serious market pressures.

Speed and focus were very much stimulated in all MuCh-QFD sessions. To a degree this could be attributed to the use of GroupSystems™, and it also had to do with the level of structure of the session, its internal logic and its focus on outputs and decisions. Communication between stakeholder perspectives (customers, channels and competitive positioning) was promoted most notably by the MuCh-QFD session in cases E: online billing and F: telecom bundle. In those sessions channel partner discussions really provided added value (which was covered more automatically by intermediaries in the other cases). In case D, offer registration, the end-customer perspective was enhanced, but intermediary wishes were somewhat suppressed in the MuCh-QFD session. One stakeholder perspective was not integrated to the extent that we expected it to be: in all cases the value IT/operations participants added to the service definition was limited. Although their input was not useless, they tended to focused on constraints, technical possibilities, design issues and implementation. At the level of the session, most service decisions did not require in-depth operational or technical expertise. (We expect this expertise to become more important when the design will be worked out in greater detail, although this is based on participant comments, not on our own observations).

In the FE sessions, the design process was only customer-oriented when the primary goal of the e-service was to improve service to the customer (case E: online billing, and partly cases: absence management and F: telecom bundle). General speaking, the degree of customer orientation was lower than in the MuCh-QFD sessions. In case C: intermediary portal, it was the 'traveler from the future' (see case description in section 7.1.1) who strongly promoted customer orientation, although this was reflected in the output (see also section 7.1.3) rather than the process. Channel value and coherence were discussed less than in the MuCh-QFD sessions of cases A: absence management, B: insurer portal, C: intermediary portal, E: online billing and F: telecom bundle. In case D: offer registration, the participants extensively discussed cooperation between the e-service and the intermediary, and to a certain extent lost faith in the e-service. Whether this was justified or not, the result was that limited progress was made with regard to defining a multi-channel service. Discussions on channel synergy were not evoked explicitly by the FE sessions. Some of the items did come up (though not more than during the MuCh-QFD sessions) when the discussion focused in possibilities to save costs or improve the use of existing data. The FE sessions of cases B: insurer portal, D: offer registration, and F: telecom bundle, focused mainly on the supplier's point of view. Competitors were hardly discussed in any detail during FE sessions, except for cases D: offer registration, and F: telecom bundle, where as a consequence of doubts about marketing strategy the group looked for ideas in what competitors were doing. In general, FE sessions evoked more marketing strategy discussions than MuCh-QFD sessions. These discussions took place during agenda parts I, when e-service objectives were discussed (cases C: intermediary portal, E: online billing, F: telecom bundle), and during part III, when detailing the e-service based on objectives and functions (all cases). All three topics of part III (e-service slogan, online solutions, invest or not) raise the question 'why are we doing this?' Especially the question whether the e-service warrants investment triggers discussions on expected market success and remaining bottlenecks. However, extensive discussions on marketing strategy are not always a sign of strength. For example, wlthough there was a great deal of discussion on marketing strategy in case F: telecom bundle, was mainly due to uncertainties regarding e-service objectives and priorities, which hampered rapid progress, especially in part III of the FE agenda.

In FE sessions speed and focus are high, very similar to MuCh-QFD sessions. This is caused largely by similarities in formats, procedures and the use of GroupSystems™. Communication between stakeholder perspectives regarding design decisions was limited, with the exception of case C: intermediary portal, where customer, multi-channel, IT and marketing strategy issues were all integrated into the service concept. However, this integration was performed by the illustrious 'traveler from the future'. The disadvantage of his otherwise valuable contribution was that it turned the second part of the session into a kind of 'one man show'. In general, IT/operations integration into service definition was limited, just like in the MuCh-QFD sessions. Interestingly enough, if we compare the various cases, we see that during the FE sessions extensive discussions about objectives and marketing positioning took place. However, the results of these discussions were limited in terms of integrating multiple perspectives: in practice they tended to focus primarily on the interests of suppliers.

We conclude this evaluation by summarizing the contributions to our design requirements. With regard to customer orientation the intake is especially useful for e-services that are not already inherently customer-oriented (as in case E: online billing, which had customer satisfaction as main objective). E-services initiated on the basis of supplier, technology, competitor or legislature considerations benefit from the intake because it increases the emphasis on customer orientation. As far as the sessions are concerned, FE does not lead to customer orientation for the latter type of e-services, whereas MuCh-QFD sessions do. Our second requirement, channel coherence, is strongly stimulated in MuCh-QFD sessions, in contrast to FE sessions. FE sessions only addressed channel coherence issues when the intermediary channel is an important part of the e-service concept from the start. And even then, translating channel coherence into e-service definition decisions is stimulated to a lesser extent. Thirdly, channel synergy was less noticeably stimulated by both MuCh-QFD and FE sessions; most participants indicated they needed additional information before being able to consider ways to realize inter-channel synergy. With regard to competitive positioning, our fourth requirement, MuCh-QFD sessions stimulated detailed comparisons with competitor e-services (which were virtually absent from FE sessions), and FE sessions stimulated marketing strategy discussions ('why are we doing this and how should we do it?'), which in term were virtually absent from the MuCh-QFD sessions. Both MuCh-QFD and FE sessions had high levels of speed and focus (requirements five and six), due to similarities in process conditions. This high performance was partly caused by the use of GroupSystems™ and partly by the degree of structure of the session, its internal logic and its focus on outputs and decisions. Also, the information provided during the intake contributed to the speed and focus of the sessions. With regard to requirement seven, communication between stakeholder perspectives, the MuCh-QFD and FE sessions had quite different effects. In the FE sessions, the results of the discussions on objectives and marketing positioning were limited in terms of integrating multiple perspectives: in practice these discussions tended to focus primarily on supplier interests. The MuCh-QFD sessions, on the other hand, did stimulate communication between stakeholder perspectives, most notably for e-service ideas whose initial goal was not to incorporate the priorities of other channels and/or customers.

7.1.3   Session outputs

In the previous section we discussed the design processes for MuCh-QFD versus FE. In this section and the next we focus on session outputs. It is not possible to represent and discuss the results of all sessions in detail, for reasons of space and confidentiality. In this section we illustrate the outputs for both sessions with outputs from case C: intermediary portal (which is the case with the least amount of confidentiality constraints). In the next section (7.1.4) we evaluate session outputs in relation to our design outcome quality requirements across all cases and sessions.

The output of both sessions consists of:
1.   Information captured via GroupSystems™:
      brainstorm results; clustering results (for example the main customer needs in MuCh-QFD, or the main objectives in FE), and priorities as scored by the participants
2.   Flip-overs from breakout exercises
3.   Summary of outputs in service definition templates in MS Excel

The first type of output: information captured via GroupSystems™, is different for the MuCh-QFD and FE session. These differences are further analyzed in Appendix H and the next section. What is similar is that both sessions use the same approaches and tools via GroupSystems™: both cases included the sequence of first exploration (on objectives, or customer needs, or functions) and then clustering of the previous results into the main categories via a group discussion and centralized category definition aided by the facilitators.

The second type of output: flip-charts from breakout exercises, are also different for MuCh-QFD versus FE. Agenda part II of MuCh-QFD, define e-service matrix, uses a flip-chart on which a matrix is drawn that contains the customer needs and Web functions (as shown in Figure 7-1). Then the 9's and 3's are filled in, and the e-service slogan is drawn in the top left corner. Below the matrix there is a second flip-chart, with the same columns as the e-service matrix. All remarks on possible online tools for fulfilling functions, and relations between two different functions and design constraints, are collected in those columns. In agenda part III, tasks of other channels, the Top 3 support needs from other channels and the Top 3 win-win options between the e-service and other channels are written on a new flip-over sheet. This is a reference point for participants when the original e-service matrix is expanded with multi-channel related customer needs (which create additional rows) and multi-channel service functions (which create additional columns), see also Figure 7-1. In direct continuation of the previous tasks, part IV of MuCh-QFD: competitive position, is briefly supported via a group exercise supported by a flip-chart, by capturing the main strengths and weaknesses that participants perceive in relation to competing e-service offers. This helps participants to form an opinion before they generate their competitive position scores via GroupSystems™. In FE sessions flip-charts are used only for agenda part III: solutions and new e-service. In FE sessions there is only one breakout exercise, compared to two in the MuCh-QFD sessions, although the overall breakout lengths are similar (50 and 60 minutes, respectively). In FE sessions, a first flip-chart is used to highlight the three sub-tasks (define e-service slogan, define main online solutions, indicate why to invest or not) and capture the results for part III. And a second and sometimes third and fourth flip-charts are used by the group to draw either Web pages, or task hierarchies or a list of main solutions (groups are free to choose in this task how they create their outputs). The web outputs provide a reference point for the participants when they discuss whether the e-service merits investments or not.

In the remainder of this section we illustrate the third and main type of output: the summary of the results per session in MS Excel. Figure 7-1 provides an example of a MuCh-QFD service matrix. It contains elements similar to the illustrations we provided in 4.2, but all the main output items are put together here. At the bottom, this figure adds solution remarks that participants made with regard to each function, in the top left corner it also adds strengths and weaknesses as seen by the participants, and just below the strengths and weaknesses it states the service slogan. And as also explained in 4.2 it contains customer needs and functions for the e-service as well as for the multi-channel services, and it helps connect discussions on solutions and functionality to customer priorities. It is likely that certain customer needs or functions as defined in the left and upper parts of the service matrix may be rather abstract for outsiders. Nevertheless, each of these concepts summarizes a number of specific brainstorm outputs that anyone can read back in the GroupSystems™ output after the session.

In Figure 7-2 the FE output summary is illustrated. At the top a list of prioritized objectives is presented along with indications as to how the new e-service scores in relation to the importance of the objective (Importance Performance Analysis, see 5.3). The top left corner also shows the service slogan that summarizes the service concept. The next table is a summary of the main functions and the priority scores provided by the participants. Further below the core Internet solutions are presented, including remarks on: what is feasible or not, which functions work best and why, etc. The table at the bottom provides reasons in favor of and against investing in a particular e-service concept. This provides input for the next steps in the design and decision-making process.

Strengths
Price/quality
2 options: customer or IM initiative
The time is right

Weaknesses
Politics: Finding partners; building scale
Effective PR: how?

Slogan= 24x7 your personal advisor at home

Customer needs	Weights	FAQ	Mini-calculation with incentives	Process overview	Calculation function and comparison	Call-back function	Links to and information about providers	Product information	Recommendation complementary products	Cancellation service	Personal interview	Price incentive	Active migration customers	New	Present	Direct writers	Banks
1. Price information and price benefit	9		9		9									8	8	7	6
2. Comparing premiums and conditions	8,5				9									9	9	4	4
3. Quality product	7,5	3		3		3		3						7	6	7	7
4. Quality provider	7,25						9	9		3				7	6	7	7
5. Tailor-made advice	7,25	3			3				9		9			6	7	6	7
6. Product information	7							9						8	7	8	8
7. Insight into process	6,75	3		9						3				6	6	6	7
8. Insight into legal process	4,25	3		9		9	3			3				6	4	5	6
(9. User-friendliness)	8																
10. Offer me channel that I am used to	7,75					3					9	9	9	6	7	5	6
11. Need for other products	6					3			9		9			7	7	8	8
Importance:		80	81	122	179	135	78	156	119	66	189	70	70	497	423	441	461

Annotations:

Function	Notes
Mini-calculation with incentives	Flow chart
Process overview	(Small print)
Calculation function and comparison	Maximize sales; Use known customer info; Package deal
Product information	Online ab offer & offline xy; Check boxes issues & products; Show package deal
Cancellation service	Say: we handle it; (Standard protocols)
Personal interview	Say: online= fast & easy; Mail: online xyz, visit abc
Price incentive	Price incentive
Active migration customers	Other brand?; Marketing & PR

Figure 7-1 MuCh-QFD output illustration (case C: IM portal; importance = ∑weights x scores)

Slogan= Best of both worlds: your trusted advisor 24x7 and cheap

Objectives	Weights	Concept score	Difference score
1. Accessible, user-friendly and interactive	8,25	9	0,75
2. Good price/quality	7,75	7,5	-0,25
3. Save costs	7	8,5	1,5
4. Improve competitive position	6	5,25	-0,75
5. Central customer folder	5,75	9,25	3,5
6. Offer product information	4,75	7,25	2,5
		Total score:	7,25

Functions	
1. Compare product and prices	9.25
2. Advice tool	8.75
3. Value added by IM	6.75
4. Manage active insurances for customer (IM & customer)	6.75
5. Product information	6.25
6. Enable IM-specific features and management	4.00

Core Solutions
1. Compare offer to price of customer's existing policies
- be clear that policy may run 3 more years, and offer to change it at that time
2. Start and stop immediately
- customer can stop a policy any day
- confirm everything per mail and send policy immediately
3. Report, settle and track damages
4. Added value IM
- show online what you did for the customer this past year
- info on previous contacts
- proactively offer advice (on other products, insurers etc)
- explain way of working and offer contact options
(background discussion: customer perceive more activity, but a lot is automated)

Invest (y/n)?
1. In case of mandate model:
- IT providers and IM's must move. IM's are starting to come along
2. If outside mandate model = Pool of insurers:
- at least 3 to 4 insurers need to cooperate and invest in infrastructure

Figure 7-2 FE output illustration (case C: intermediary portal)

For both session types the output summaries provide a stepping-stone for communicating the results to others and help people move towards the development phase. However, the summaries tell only half the story. It is important to know that there is a basis underlying each summary. This basis consist of 1) further details (for example, by the third FE function 'added value agent' we summarize the eight sub-functions from our brainstorm that are in our GroupSystems™ session output report), and 2) shared notions between stakeholders (for example, 'we discussed the objectives and these are the main ones that we see').

In conclusion, if we compare the summaries of MuCh-QFD and FE sessions, we can see that the results of MuCh-QFD tasks are more integrally connected (for example functions as well as scores on competitor e-services are directly linked to customer needs) than the FE results. Also, regarding the content of the outputs, MuCh-QFD results show more attention to customer needs, multi-channel tasks, and competitive e-service evaluations than FE results. This is true for all the cases we observed (no FE group spontaneously listed customer objective instead of the more general objectives or a comparison with competing e-service offers as part of the evaluation whether to invest or not). The combined effect of these two MuCh-QFD characteristics (format and content) is what creates a high performance in terms of our design requirements customer orientation, channel coherence, competitor focus and communication between stakeholders. This is confirmed very directly by the participant opinions analyzed in sections 7.1.5 and 7.1.6.

7.1.4   Session output analysis

In this section we present the main findings of our output analysis of the MuCh-QFD and Fundamental Engineering sessions. More details can be found in Appendix H, which shows the results the participants generated for the MuCh-QFD and FE sessions of all six cases in terms of customer orientation, channel coherence and synergy, and competitive positioning.

With regard to customer orientation, it is clear that in the MuCh-QFD approach the focus is on customer requirements, while more general business objectives are being emphasized in the FE sessions, and any focus on customer requirements is more accidental. This does not mean that more general issues were not discussed in the MuCh-QFD sessions, but they were invariably looked at from the customer's point of view. When process management issues were discussed in MuCh-QFD sessions, the participants were always aware that users need to understand which steps have to be taken or what relevant information could be retrieved. In case C, the intermediary portal, one participant in the FE session very actively promoted looking at the customer's point of view and considering customer requirements. Moreover, he had been commercially active with a similar service for about 2 years and provided many useful insights into customer behavior and how customers had responded to different services. In section 7.1.2 we called him 'the traveler from the future'. In the FE session of case E there was also a strong customer orientation: this was caused by the fact that customer orientation was a very central element of the service concept in question. All the other FE-sessions focused on supplier interests. Service requirements were defined preliminarily from a supplier's rather than a user's perspective.

Customer orientation is also reflected in the functionality as defined in the MuCh-QFD sessions versus the FE sessions. In the MuCh-QFD sessions a greater number of references was made to tools that are relevant to customers, such as notifications, direct links from intermediary Internet sites to those of insurers (case A: absence management), support and advice tools (case B: insurer portal, and case F: telecom bundle), self management, support tools and support for personal contact (case D: offer registration, and case F: telecom bundle), while the FE sessions focused more on back office processes and visualization (case A: absence management), smart forms, reusing data, value chain integration (case B: insurer portal), functionality focused on control of liability (case D: offer registration) and showing bundle benefits, promoting cross-selling and up-selling (case F: telecom bundle). Again, there was a greater degree of similarity between the MuCh-QFD- and FE sessions in case C: intermediary portal, and E: online billing. Some functionalities, like product and price comparison tools in the insurance case and one-stop-shopping functionality, analysis and advice were mentioned in both sessions of these two cases. Overall, if we look at the degree of customer orientation that is translated into functional specifications, it appears that MuCh-QFD has limited added value when cases are very customer-oriented by their nature (due to participants, see case C: intermediary portal, or the service idea, case E: online billing), whereas in other instances (see the other cases) it has.

The second requirement, channel coherence, was discussed in great detail in the MuCh-QFD sessions and incidentally addressed in the FE sessions, with the exception of case C, intermediary portal, were channel coherence was discussed very explicitly. In the service definition of all the insurance cases the focus was on combining Internet-based services with the added value of personal contact provided by intermediaries. In the MuCh-QFD sessions this was reinforced in discussions on channel coherence. In the FE sessions, however, this focus, although implicitly central in every case, was lost. The FE sessions focused more heavily on process-related benefits and on workflow management, which resulted in suggestions being made regarding the visualization of the process, the opportunity to use electronic forms to generate quotations or (cost-saving) customer self-management. In the MuCh-QFD sessions many win-win ideas for multi-channeling were discussed. In all MuCh-QFD sessions the combination of e-services with face-to-face interaction was suggested. Service personalization was a driver, not only with regard to the use of e-services, but also with regard to face-to-face meetings.

Channel synergy, in other words reusing assets across the various channels, is an important issue. In all cases and in almost all of the sessions the participants looked at back office and system integration issues. Comparatively little time was spent discussing commercially clever ways to reuse investments or information, or ways to optimize the timing of customer contacts and process integration. From the feedback provided by the participants we conclude that they thought it was too early in the design process to identify detailed synergy opportunities. In the MuCh-QFD session of the absence management case (A) the emphasis was on easily achievable benefits like the electronic exchange of reports to the actors involved. In the FE session of the same case as well as the online billing case (E), possibilities to eliminate paper-based processes were discussed. In most of the other cases integration within the value chain was taken for granted. The discussion focused on the extent to which integration was desirable. Identifying financial benefits was considered 'still impossible, or at least immature'. This meant that the discussions were fairly general in nature.

With regard to competitive positioning, the competitor benchmark tool as used in MuCh-QFD proved to be very valuable. Although we should not take the exact scores too seriously, the differences in score do provide an indication about how well competitors are doing. Differences between what competitors had to offer, the existing service and the new service concept could be made explicit at a very detailed level. The new service concepts scored better than the existing and competing services in all cases for the MuCh-QFD sessions. Although the scores were more or less artificial, they helped the design team to focus on the strong points and improve the weaknesses. In the FE sessions the discussions on the competitive positioning of the services were more random. Interlocked relationships, not only between suppliers and customers, but also within the value chain, were discussed (case A: absence management, B: insurer portal, case C: intermediary portal). In the other cases the key issues were cross-selling and branding or higher general competitiveness. Overall one could say that MuCh-QFD session outputs more directly provide competitor comparisons, and FE sessions more broadly address marketing strategy questions. This qualitative finding is confirmed by the measurement instrument scores in sections 7.2.2 and 7.2.3.

In summary, customer orientation was more reflected in MuCh-QFD session functionality definitions than in FE session functionality definitions. An exception to this rule was formed by case C: intermediary portal, with one very customer-oriented 'traveler from the future' and case E: online billing, with an e-service idea which was inherently very customer-oriented. In the other cases FE sessions focused more on value chain integration, smart forms, reusing data, promoting products or suppliers and stimulating cross-selling and up-selling. The second requirement, channel coherence was discussed in great detail in the MuCh-QFD sessions and incidentally addressed in the FE sessions. This difference occurred despite the fact that in all the insurance cases the focus was on combining Internet-based services with the added value of personal contact with intermediaries. The FE sessions focused more heavily on process-related benefits, the opportunity to use electronic forms for generating quotes or (cost-saving) customer self-management. An exception was case C: intermediary portal, were channel coherence was discussed very explicitly. MuCh-QFD and FE did not generate many differences in relation to the third requirement, channel synergy. In virtually all of the sessions the participants looked at back office and system integration issues. Comparatively little time was spent discussing commercially clever ways to reuse investments, information or other assets. Participants thought it was too early in the design process to identify detailed synergy opportunities. Regarding competitive positioning, our fourth requirement, our results show that MuCh-QFD session outputs more directly provide competitor comparisons, and FE sessions more broadly address marketing strategy questions.

To conclude we can say that our qualitative analysis of the output confirms the process analyses (section 7.1.2) and measurements (sections 7.2.2 and 7.2.3). Furthermore, we conclude that generally speaking the degree of *process* attention to outcome-related design requirements (customer orientation to competitive positioning) correlates well with the degree of *output* compliance to those design requirements. (Case C again is an exception, since its output was not generated via a group process, but was based on the input of one 'traveler from the future'.)

7.1.5 Agenda item preferences stated by participants

This section addresses participant opinions. Participants were asked to express which agenda items they thought were the most and least useful (questions 9 and 10 in the questionnaire, see Appendix E), with multiple responses allowed, up to a maximum of 5 agenda items per question. Also, they were asked to explain their scores. We have summarized the MuCh-QFD scores in Table 7-2.

Table 7-2 Number of times an agenda item was scored 'most' or 'least' useful by MuCh-QFD participants (maximum of 5 responses allowed)

MuCh-QFD session agenda	Most useful	Least useful
Part I: Customer needs and Internet functions		
- Identify, cluster and prioritize customer needs	20	1
- Identify and cluster Internet functions	10	5
Part II: Define e-service		
- Evaluate functions with regard to needs and create an e-service matrix	11	3
- Define service slogan that summarizes proposition	2	17
- Discuss solutions and constraints for functions	11	6
Part III: Tasks of other channels		
- Check the desired support from other channels	10	5
- Check win-win options between e-service and other channels	15	3
- Extend matrix with needs and functions related to other channels	4	7
Part IV: Competitive position		
- Score existing, competitors' and new site on customer needs & discuss results	14	7
Number of MuCh-QFD participants:	**(n=24)**	**(n=24)**

This table indicates that overall MuCh-QFD participants were more positive than negative about agenda usefulness. The first activity, identifying prioritizing customer needs, is valued as one of the most useful agenda items by 20 of the 24 participants. Their general explanation was that they see customer needs as the starting point and as a basis for the steps that follow. The service slogan definition, in part II of the agenda, was found 'least useful' (explicitly stated by 17 of the 24 participants). The general opinion was that this element did not contribute directly to the service definition. Also, people associated it with promoting and advertising the service, which many felt would be inappropriate at this particular stage in the service introduction process. With regard to the multi-channel activities it is interesting to note that the participants were relatively positive about exploring the other channel support needs of customers, and looking for win-win opportunities between channels, although they were less positive about translating this to multi-channel service matrix consequences (7 votes 'least useful' versus 4 votes 'most useful'). This again reflects the difference between corporate versus SME participants: 5 of the 7 'least useful' votes come from the three SME cases (cases A: absence management, C: intermediary portal, D: offer registration), and 4 of the 4 'most useful' votes come from the corporate cases (cases B: insurance portal, E: online billing, F: telecom bundle). Since managing multi-channel service is more complex for corporate firms, the service matrix adds more value for them in their ability to manage multi-channel service coherence. (The underlying theoretical assumption is that inclusion in the service matrix enhances manageability. This assumption is confirmed by the findings in the next section.) SME participants indicated that the service matrix extension did not add much, which confirms our observations in section 7.1.1 that multi-channel service execution and management are not key problems to them.

Next, Table 7-3 shows results for the FE sessions. As in the MuCh-QFD group, participants are generally more positive than negative about the usefulness of the various agenda items. This time two items stand out as being considered particularly useful by the participants. Firstly, identifying objectives (21 out of 25) is considered important because it is the basis of the e-service; it answers the 'why' question. Secondly, identifying Internet functions (19 out of 25) is valued because it makes the e-service idea more concrete.

Table 7-3 Number of times an agenda item was scored 'most' or 'least' useful by FE participants (maximum of 5 responses allowed)

FE session agenda	Most useful	Least useful
Part I: Objectives		
- Identify and cluster objectives	21	2
- Prioritize objectives	12	5
Part II: Functions		
- Identify and cluster functions	19	1
- Prioritize e-service functionality clusters	11	2
Part III: Solutions and new e-service		
- Define service slogan that summarizes proposition	5	12
- Develop an e-service based on outcomes part I and II	11	5
- Indicate why to invest or not	10	10
Part IV: Assess the new e-service		
- Score the e-service on the objectives	14	6
Number of FE participants:	(n=25)	(n=25)

Similar to what we saw in MuCh-QFD sessions, FE participants value the service slogan least (12 out of 25 'least useful' scores versus 5 'most useful' scores). Again participants indicate that it does not add much, and/or that it comes too early. Nevertheless, five participants – none of them SME's – are positive. The explanations that are provided are that 'it brings the service to life', that it 'helps guard the core value of the e-service' and that it presents the e-service 'at a glance'. Finally, we want to discuss the scores on the activity 'indicate why to invest or not', which is 10 times voted 'most useful' and 10 times 'least useful'. The latter group indicates that 'it comes too early' and is too uncertain to contribute much. The first group states that they found it useful to discuss this activity, since it provides an indication for chances of success for the e-service idea and helps identify potential bottlenecks.

If we compare the scores for the session types, there are a few things that stand out. One is that participants seem to find it easy to adopt the design method that is presented to them as 'the right way to do it'. As such, MuCh-QFD participants find it logical to view customer needs as the basis for the e-service definition, whereas FE participants consider objectives as the logical basis. Secondly, although the two sessions are different, there are a few comparisons we can make. Two activities are present in both sessions: 'identify Internet functions' and 'define service slogan'. For both activities, FE participants are more positive and less negative about their relative usefulness than MuCh-QFD participants. With regard to the Internet functions, this can probably be explained to some extent by the fact that MuCh-QFD sessions provide a broader scope: defining multi-channel servicing. For FE sessions the Internet often remains the center of attention, which could explain why Internet function definition is seen as more important. With regard to service slogan definition, it is less easy to find an explanation. It could be that service slogan definition better fits the FE approach, or that MuCh-QFD activities are generally more valuable than FE activities, which makes the contrast within a session with service slogan definition bigger. Another reason may be that during agenda part II of the MuCh-QFD session the session focuses very much on filling in and discussing the service matrix. By contrast, service definition occurs at the start of part III of the FE session, which is less structured (service slogan definition has actually been put there as a consequence of our pre-tests, which indicated that part III of FE needed a 'stepping stone' to get started). Hence, we think that in the FE session it adds more value as an anchor point, due to its relative position in the session.

Finally, we want to emphasize again that all the participants were generally speaking positive about the usefulness of the agenda items. Also, the participants often explicitly said that just because they have indicated that they found a particular item 'least useful', that did not mean that they considered the item useless.

7.1.6   Feedback from debriefings

For each case we had a debriefing with participants about one month after the design sessions. In these meetings FE and MuCh-QFD participants were present together. The methods and results of both sessions (FE and MuCh-QFD) were discussed here. Also, the results of the two sessions were compared and we provided some recommendations for next steps per e-service case, based on the combined output of both sessions per case. Finally, we held a brief evaluation discussion, centered on six evaluation questions. We first asked participants to raise their hands to give us their answers (yes, no or neutral), after which we asked each of the participants to explain his or her answer.

The debriefing questions were geared more towards our MuCh-QFD approach than towards the FE sessions. We explained that FE has been around for a long time, and is often used as a tool by one or two individual designers to help them expand the design space and avoid 'jumping to solutions', and that MuCh-QFD originated from corporate settings where multiple departments have to cooperate in design and innovation. We emphasized that, although both approaches are valuable in their own right, we expected that MuCh-QFD would be more suitable when it comes to defining e-services with multiple parties involved. This was the reason we wanted to compare the effectiveness of the two approaches, and also why we were particularly interested in the effectiveness of the MuCh-QFD approach.

In Figure 7-3 we show participant answers. The number of participants present at the debriefings, n = 28, is lower than the number of people (49) who took part in the e-service definition sessions. Because people were asked to attend the debriefings on a strictly voluntary basis, and some of the participants were unable to come, not everybody did attend. This self-selection of participants might bias the data, so the results must be interpreted with some care. For case D, the participants of which were the least committed of all that took part, only 2 participants (from MuCh-QFD) were present. In all other cases, 4 to 7 of the participants were present, relatively equally divided between the FE and MuCh-QFD sessions. Only in case A there was a relative imbalance, with 4 MuCh-QFD participants versus 2 FE participants. In total, there were 17 participants from MuCh-QFD sessions and 11 from FE sessions. Statistically, the answers provided by the participants of the two sessions are not significantly different, which is why the answers of both types of sessions have been clustered in Figure 7-3.

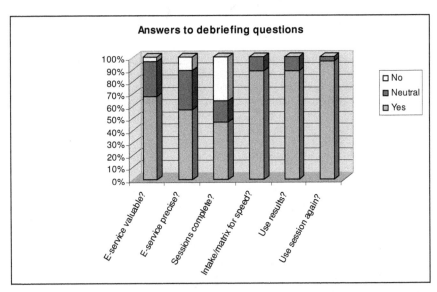

Figure 7-3 Answers to debriefing questions from MuCh-QFD (n=17) and FE (n=11) participants

Overall, participants are quite positive about the sessions and results. The six debriefing questions that we asked were:
1. Do you think the e-service definition resulting from the sessions is valuable?
2. Did the sessions help make the e-service or its positioning more precise?
3. Do you think that the sessions are complete in identifying opportunities or constraints for this e-service?
4. Do you think the intake and/or service matrix discussions with stakeholders help speed up service innovations?

5. Are you going to use the results?
6. Would you consider using one of these sessions (plus intake) again in the future?

As mentioned earlier, at the debriefing the participants were generally speaking positive about the sessions. In the following section, we discuss the participants' responses participant to the questions we asked them at the debriefing. With regard to question 1: is resulting e-service definition valuable, the participants state that it has generated a relatively complete image. However, elements of criticism are that the functions are not very detailed, and that the output is more a general e-service concept than a detailed design of an e-service.

Opinions on the second question: e-service or positioning more precise, also vary. Positive participants state that a good start has been made and some MuCh-QFD participants specifically state that they value the method's attention to customer needs. On the other hand, some of the more critical participants state that more time is needed to complete this effort.

Participants are least positive regarding question 3: completeness of the sessions. In the discussion it turned out that the question raised two issues for participants. Firstly, in relation to the time used, participants are generally positive. Some state they are very much impressed by the amount of progress and output. Also, several participants said repeatedly that that doing this at the start of the new service development process really helps speed up the process and address all the main items. This confirms that our approach, which we developed to professionalize the initiation phase of e-service definition while supporting a fast progression through the definition phase, has fulfilled that purpose, at least according to several of the participants. The second issue is that most participants agree that the output is not complete. It does help to raise issues that were new (and relevant) to the idea owners regarding, for example, the extent to which customers value price information, or the multi-channel behavior of customers, or competitive positioning issues. But for all cases there is still work that remains to be done (for example: address how to implement the e-service in the back-office, refine the business case, define Internet pages and navigation structure, further investigate competitor e-services and positioning). In the terminology we used in chapter 4: many tasks from the development phase still need to be performed before the design is ready for implementation.

In response to the final three questions, participants are relatively positive. Question 4 addresses usefulness of the intake and the service matrix for speeding up innovation. The intake is considered valuable for generating focus and avoiding discussions about scope and goals of the sessions. The service matrix is thought to be particularly valuable for two things: 1) to communicate and justify the e-service to others and help structure future discussions, and 2) to help prioritize investments and development efforts. This second aspect was most strongly welcomed by managers. They valued the direct link between each function and how important it is for customers. One of them (a manager from case E: online billing) stated that it was a shame that they were only now using it for their second service release, because one look at the service matrix already showed him they had been over-investing in one function during the first release.

Overall, three FE participants from three cases spontaneously indicated that they particularly liked the service matrix. In addition, they expressed their disappointment for not have been in the MuCh-QFD session, mainly because of the e-service matrix. The matrix is perceived to provide a clearer summary of the e-service, the priorities and the reasons behind the priorities.

When discussing question 5, using session results, the general opinion is that the output helps facilitate future communication and decisions and generate commitment. At the same time it is clear in all cases that this output does not 'settle' the issues. It is expected that multiple rounds of discussions will be needed with other people involved. In the corporate cases (B: insurer portal, E: online billing, F: telecom bundle) because 'these things simply require some rehearsal for people' and in the multi-actor SME cases (A: absence management, C: intermediary portal, D: offer registration) because the task of creating critical mass with enough other firms to co-create and co-produce the e-service had not yet been completed (which is a political and entrepreneurial challenge more than a service definition challenge). The participants of case C, intermediary portal, particularly indicated that the functions as such were not very innovative and that they were also not the key issue for the success for this e-service: the key issue was gaining momentum and adoption in the market. Across all cases, most participants indicate that even without these sessions they would have continued with the e-service, but that they were very helpful (customer priorities, service matrix, competitor positioning, investment hurdles, and multi-channel issues).

The previous remarks also help explain the positive responses to question 6, whether participants consider using our sessions for future services. Furthermore, participants refer to the design process speed, focus and the way the service matrix support follow up communication and decision making. Interestingly enough, many participants automatically focus on the MuCh-QFD approach as being the logical candidate. Only one of the FE participants stated that she thought it does not matter which method is used, as long as people discuss the e-service. However, from the feedback we received we conclude that her opinion was not shared by the other participants.

As a summary of the debriefings, participants indicate to have been surprised by the speed and amount output of the sessions (this is true for FE as well as the MuCh-QFD sessions). On the downside, several people indicated that the e-service definition is not detailed enough. Issues that were mentioned open are: refining the business case, defining Internet pages and navigation structure, further investigating competitor e-services and positioning, gaining momentum in the market and involving other parties. Overall, MuCh-QFD is considered more useful, mainly because of two points. Firstly, participants value the service matrix for future decisions and communication (especially managers were positive about this). Some FE participants indicated that they were disappointed that the service matrix had not been used in their session. Secondly, in the three corporate cases (B: insurer portal, E: online billing, F: telecom bundle) the explicit attention to multi-channel issues was appreciated and it was said this raised important (organizational and marketing) issues that otherwise would have been overlooked. Although competitor positioning discussions in MuCh-QFD were thought to have some use, they did not particularly stand out in the debriefings.

Several participants thought that it would be good to repeat the competitor scan in more detail afterwards, with more accuracy than was possible in the sessions. They indicated that they felt that an increased level of detail would also contribute more to e-service refinement and competitive positioning. The main aim of the competitor discussion was to help assess future market success, which is similar to the goal of the investment readiness discussion in part III of FE sessions. In a similar fashion this FE activity was considered to be valuable, although it did not stand out as far as the participants were concerned.

7.2   Quantitative results

In this section we discuss the backgrounds of our participants (section 7.2.1) we discuss the results from the participant questionnaires (section 7.2.2), and the observers (section 7.2.3). These results will also be used to draw conclusions on our hypotheses. Finally, we check for disturbing factors (section 7.2.4).

7.2.1   Description of business participants on the basis of background variables

The participants in our field experiment represent the following functions: fourteen directors, eight IT/operations professionals, three line managers, twenty marketers and four others. On average, 63% of them indicate that they have previously been involved in service innovations and 47% of them have previously participated in sessions in which new services were defined. Nine participants (18%) were idea owners, initiating the e-service idea and preparing the sessions with the researchers involved. Most of the other participants came from the networks of the idea owners. For practical reasons six of the idea owners participated in the MuCh-QFD sessions, and three in the FE sessions.

All background variables were comparable in the two groups, with one exception. Regarding one background variable we found significant difference between the MuCh-QFD and FE group (two-tailed t-test; t=2.52; df=45; p<0.05): in MuCh-QFD sessions participants had on average 15 years of experience in the sector, in FE sessions the average was 10 years. Subsequently, we tested if years of experience had a significant impact on any of the dependent variables (the requirements customer orientation to stakeholder communication), which was not the case. Hence, we conclude that none of the background variables we described and tested appears to represent a disturbing factor with regard to our case findings and measurements.

7.2.2   Questionnaire results

Apart from the initial background questions, the questionnaire consists of two parts: questions that measure subjective evaluations and questions that measure more objective facts. The latter questions are based on the measurement constructs discussed above. We first discuss the subjective evaluations. Possible interactions between these questions and our measurement constructs are addressed in section 7.2.4.

143

The questions all generated relatively positive answers for both groups of sessions. To some extent this could help explain why we found very few significant differences between FE and MuCh-QFD groups in terms of these questions. For example, in rating the sessions, FE participants gave an average score of 7.5 out 10, whereas MuCh-QFD participants scored an average of 7.8. In general, MuCh-QFD received slightly higher scores, though not exactly on all items, but differences were not statistically significant. In summary, it appeared that statistically speaking FE and MuCh-QFD participants were equally satisfied with the sessions a) in terms of two of the three 5-point Likert scales we used on satisfaction, b) in terms of the 3-item satisfaction factor we adopted from literature and c) in terms of final grade they gave to the sessions. Interestingly enough, one of our satisfaction questions (question 5b 'In comparison with my experience with other design methods, I am satisfied with the approach taken') did return a significant difference (two-tailed t-test; equal variances not assumed; $t=2.06$; $df=34.99$; $p<0.05$). For MuCh-QFD 22 out of 24 participants are '(very) satisfied' and two are 'neutral'. For FE 15 out 24 participants are '(very) satisfied', six are 'neutral' and three are 'dissatisfied'. Contrary to more general satisfaction surveying (Wang, Po Lo et al., 2004), it appears that in the context of our experiment, the question triggers an opinion about the design methods used and not just emotional satisfaction. Thus, our participants are more positive about the approach that was adopted in the MuCh-QFD sessions compared to the FE sessions.

When participants are asked to express their satisfaction on the sessions' contribution to each of the requirements (customer orientation to stakeholder communication) there are no significant differences between FE and MuCh-QFD. Finally, one other question generated a significant difference: question 7 'Do you think you will use the session results in the future?' MuCh-QFD participants were more positive (two-tailed t-test; $t=2.53$; $df=47$; $p<0.05$): 14 out of 24 participants will 'surely' use the results in contrast with 5 out of 25 for the FE participants. On a range of 1 to 5, MuCh-QFD scores 4.3 (standard deviation = 0.9) and FE scores 3.6 (standard deviation = 1.0).

Figure 7-4 presents the results of our measurement constructs, checking for more factual observations than participant opinions. In general, MuCh-QFD outperforms FE, except for marketing strategy, but not all differences are significant. In Table 7-4 we summarize which hypotheses have been confirmed and which have not. Based on Mann-Whitney tests, hypotheses on customer orientation (Mann-Whitney $U=77$; $Z=-4.62$; $p<0.001$), channel coherence (Mann-Whitney $U=177$; $Z=-2.50$; $p=0.01$) and stakeholder communication (Mann-Whitney $U=136$; $Z=-3.37$; $p<0.001$) have been confirmed, and the others have not. In other words, those constructs show significantly higher scores for MuCh-QFD than for FE sessions.

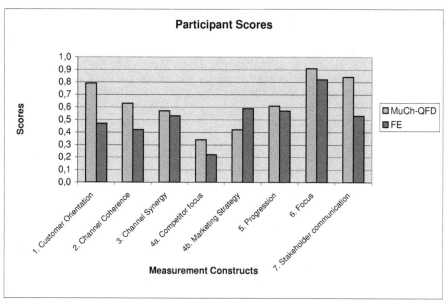

Figure 7-4 Differences between MuCh-QFD (n=24) and FE (n=25) scores

It is interesting to see that after we split up construct 4 a new significant difference has emerged: construct 4b on 'marketing strategy' shows significantly higher scores for FE sessions (Mann-Whitney U=194; Z=-2.20; p<0.05). This confirms the results of our cross-case analysis (7.1.2). These results indicate that MuCh-QFD is a relatively strict method compared to FE sessions. It prescribes how to connect customer priorities, service functions and performance levels, multi-channel needs and functions and competing offers, but it does not stimulate the more abstract marketing strategy discussions. (These discussions are not always a sign of strength, however. For example, in case F: telecom bundle, there was much strategy discussion, but this was mainly due to uncertainties regarding e-service objectives and priorities.)

Table 7-4 Confirmation of hypotheses on session effects overall

Number	Hypothesis	Confirmation
H Req1	MuCh-QFD more customer orientation than FE	Yes
H Req2	MuCh-QFD more channel coherence than FE	Yes
H Req3	MuCh-QFD more channel synergy than FE	No
H Req4	MuCh-QFD more competitive positioning than FE	No
H Req7	MuCh-QFD more stakeholder communication than FE	Yes

7.2.3 Observation results

The observation measurements generate additional insights into the differences between MuCh-QFD and FE for each agenda part. The results are shown in Figure 7-5. The four graphs display MuCh-QFD scores on the left and FE scores on the right (averaged over all sessions) and 'design outcome' quality requirements above and 'design process' quality requirements below.

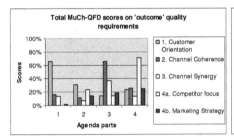 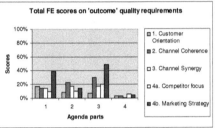

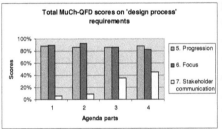 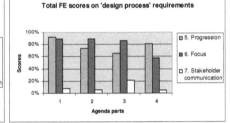

Figure 7-5 Observer scores for MuCh-QFD (n=28) and FE sessions (n=28) per agenda part

In Table 7-5 the findings in relation to our hypotheses are summarized. We start with the measurements on 'design outcome' quality requirements.

Table 7-5 Confirmation of hypotheses on agenda parts

Number	Hypothesis	Confirmation
H Part1.Req1	MuCh-QFD more customer orientation than FE in part I	Yes
H Part2.Req1	MuCh-QFD more customer orientation than FE in part II	Yes
H Part3.Req2	MuCh-QFD more channel coherence than FE in part III	Yes
H Part3.Req3	MuCh-QFD more channel synergy than FE in part III	No
H Part4.Req4	MuCh-QFD more competitive positioning than FE in part IV	No
H Part3.Req6	FE less focus than MuCh-QFD in part III	No
H Part1/4MuCh-QFD.Req7	MuCh-QFD more stakeholder communication in part IV than in part I	Yes
H Sum of Parts.Req7	MuCh-QFD more stakeholder communication than FE in all parts summed	Yes

146

In part I of the agenda, customer orientation scores relatively high (65% of the maximum score) in the MuCh-QFD sessions, which is significantly higher than in the FE sessions (Mann-Whitney U=71; Z=-5.86; p<0.001). This confirms hypothesis H Part1.Req1 (see 5.5). Regarding our new construct 4b, marketing strategy, FE sessions score significantly higher (Mann-Whitney U=163; Z=-4.55; p<0.001) in part I than MuCh-QFD sessions. However, this score is not significantly higher than it is for part IV of the MuCh-QFD agenda. This implies that discussing objectives triggers statistically similar degrees of marketing strategy discussion as is done by the explicit competitor comparisons in agenda part IV of MuCh-QFD.

In part II of the agenda, customer orientation scores significantly higher in MuCh-QFD sessions than in FE sessions (Mann-Whitney U=197; Z=-3.63; p<0.001). This confirms hypothesis H Part2.Req1. Moreover, this part II score is significantly higher than the highest FE score, which is found in in agenda part I (Mann-Whitney U=269; Z=-2.19; p<0.05). This indicates that e-service discussions via an e-service matrix (HOQ room 3 in QFD) appear to generate a higher level customer orientation in the design approach than any FE task (discussing objectives, functions, solutions or investment readiness).

In part III of the agenda channel coherence scores significantly higher for MuCh-QFD sessions than for FE sessions (Mann-Whitney U=132; Z=-4.35; p<0.001). This confirms hypothesis H Part3.Req2. Channel synergy scores higher, but not significantly. Hence hypothesis H Part3.Req3 is not confirmed. This confirms our cross-case analyses results of 7.1.2. Regarding our new construct 4b, marketing strategy, FE sessions score significantly higher (Mann-Whitney U=192; Z=-3.45; p<0.001) than MuCh-QFD sessions. Moreover, the part III marketing strategy score is significantly higher (Mann-Whitney U=225; Z=-2.91; p<0.01) than the highest MuCh-QFD score (occurring in part IV). This implies that the part III topics of FE sessions (e-service slogan, online solutions, invest or not) trigger more marketing strategy discussions than competing e-service initiatives as performed in part IV of MuCh-QFD sessions.

In part IV of the agenda construct 4a, competitor focus, has a relatively high score (70%) for MuCh-QFD sessions, which is significantly higher (Mann-Whitney U=34; Z=-6.22; p<0.001) than for FE sessions. Although this does not directly confirm hypothesis H Part4.Req4, it does point in the same direction. One could say that the assumption that more attention is paid to competitors in MuCh-QFD sessions is confirmed. In general all outcome requirements score relatively low in part IV for FE: according to our measurements relatively little happens here in relation to our requirements. This confirms our analyses in 7.1.2: part IV of FE is important in that a) it provides a sense of 'closure' to participants, and b) it enables reflection on remaining issues. Most of these issues (for example: ease of use, finding partners/coalitions in the market, internal organization or process issues, business case uncertainties) may be valuable, but they are not present in our constructs.

Next we discuss the measurements on 'design process' quality requirements. In general the scores on progress and focus are relatively high and relatively consistent across agenda parts. Also, there are no significant differences between FE and MuCh-QFD scores. This confirms our general assumptions in 5.5, and implies that hypothesis H Part3.Req6, that there might be less focus in part III of FE sessions, is not confirmed. In parts I to III no significant differences can be found between FE and MuCh-QFD sessions for any of the process measurements. It is only in part IV of the agenda that there is a significant difference with regard to stakeholder communication (Mann-Whitney U=93; Z=-5.29; p<0.001). This part IV score of MuCh-QFD session is also significantly higher (Mann-Whitney U=207; Z=-3.16; p<0.01) than the highest FE score (in part III). Finally, both remaining hypotheses regarding stakeholder communication are confirmed. Firstly, hypothesis H SumofParts.Req7, that the sum of agenda parts I to IV generates a higher score for MuCh-QFD than for FE sessions is confirmed (Mann-Whitney U=172; Z=-3.65; p<0.001). Secondly, hypothesis H Part1/4MuCh-QFD.Req7, that MuCh-QFD agenda part IV scores higher than MuCh-QFD part I is confirmed (Wilcoxon test for 2 related samples, Z=-3.48; p=0.001).

7.2.4   Analysis of disturbing factors and interaction effects

In the previous sections we illustrated the statistical results regarding the primary relationships we wanted to test: the effect the design method has on performance in relation to our seven design requirements. In this section we discuss and to some extent statistically check for disturbing factors, for example, does previous session experience effect how our requirement constructs are scored, and are these effects different for FE versus MuCh-QFD sessions? Does firm size, a participant's function (marketing versus IT), etc? In this section we attempt to gauge these effects (even though we do not have enough observations for robust analyses of variance). We use analysis of variance, under the assumption that our constructs are sufficiently similar to additive scales and thus generate interval variables. We must be cautious in drawing conclusions from the analyses for three reasons. Firstly, we assume interval scales. Secondly, our number of observations is very small. Ideally, one would want at least 30 observations per cell, and even our simplest 2x2 covariance matrices cut down to 12 of 13 observations per cell or less. Thirdly, all our results are based on only six cases, which is potentially the weakest point in our experiment. One or two 'strange' sessions can have a significant effect on the statistics.

For the following variables no significant effects were found on the basis of analysis of variance with our seven design requirement constructs: participant's function (manager, project manager, IT specialist etc) [question 1c from the questionnaire]; years of working experience in the sector [question 1d]; experience in developing similar services [question 4]; opinion on method used in relation to previous experience [question 5b]; expectations of session's value for rapid service definition [question 6]; being e-service idea owner; firm size (single corporate firm session versus session with participants from several SME's); and industry background (insurance versus telecommunication). For the other variables, we describe the effects found below.

For the variable **stakeholder perspective during session** [question 2] we found a linear effect in addition to the main 'session type' effect of our field experiment. On construct 4a, competitor focus, stakeholder perspective does have a significant impact (Anova, F=5.7; p<0.01). At first sight this appears to be caused by the marketing perspective, the competitor focus mean of which is higher than average, and the operations/IT perspective, the competitor focus mean of which is lower than average (see Appendix I for statistics details). However, t-tests to check this generate no significant results, although the operations/IT perspective comes very close (two-tailed t-test, t=-1,93; df=47; p=0.06). The fact that t-tests show no differences, where Anova does, may either be caused by the unreliability of Anova with few observations per cell, or by the fact that t-tests do not compensate for the fact that part of the variance in the competitor focus scores is caused by session type. If the linear effect from the Anova is statistically valid, and the decreased scores from the operations/IT perspective and heightened scores for the marketing perspective are internally valid differences from the averages of the other perspectives, this points towards lower and heightened sensitivity for competitor issues. This is in line with common sense: operations/IT professionals mostly ignore competitive positioning details, whereas the marketing perspective is very much focused on these items. Since both perspectives are present in each session, this effect did not disturb our main experimental findings regarding the impact of session type on competitor focus.

The variable **previous participation in service definition sessions** [question 3] shows an interaction effect with session type for the construct channel synergy (Anova, F=6.71; p<0.05), see Appendix I for statistics details. We checked the covariance matrix and the raw data for potential case-dependent effects, but found no sensible explanation. Why in particular channel synergy would show this interaction effect (see also Table 7-6) we do not know. This may be a random effect (channel synergy scores were found 'moderately reliable' in our cross-measurement reliability check of 6.4.4). Since session experience was roughly equally divided across MuCh-QFD and FE, this interaction effect does not change our conclusions regarding the main experimental effect (session impact on channel synergy).

Table 7-6 Channel synergy averages: session type x previous session experience (y/n)

	MuCh-QFD	FE
Session experience	3.3 (n=12)	4.5 (n=11)
No session experience	4.6 (n=12)	3.1 (n=14)

Regarding the variable **satisfaction** [factor of questions 5a to 5c], there is an interaction effect with session type for the construct channel synergy (Anova, F=2.86; p<0.05), see also Appendix I. This effect is difficult to interpret: scatter plots provide no insight, and if we analyze the MuCh-QFD and FE groups separately, no significant Pearson correlations between channel synergy and satisfaction show up. Thus, we cannot conclude what the direction of the interaction effect is, nor can we understand it. The interaction may be a random effect. Since satisfaction was statistically equal for both MuCh-QFD and FE groups we conclude that this interaction effect has not disturbed our main experimental effects.

There also is a positive linear effect with design process focus: the more satisfaction, the higher the focus scores are (Anova, F=4.44; p<0.05). The

Pearson correlation is also significant (r=0.40; n=48; p<0.01). From our experimental design we cannot conclude what is cause and what is effect. Is higher satisfaction attributed to focus in retrospect, thus generating higher focus scores? Or have participants experienced higher levels of satisfaction when they experienced more focus? Of course, the first explanation does raise the question (which makes its logic less compelling) why satisfaction would only be attributed by participants to focus, and not to progress, stakeholder communication or even the design outcome quality requirements? Our observations during sessions (see also 7.1) show that participants were very positive about the degree of focus during our sessions: it is an explicitly appreciated characteristic of the sessions. This finding confirms the second explanation: participants became more satisfied when they experienced more focus.

Participation during intake shows a positive linear effect, in addition the primary effect of session type, on channel coherence (Anova, F=4942; p<0.01), see also Appendix I. This must be interpreted with some care, since only 11 participants were present during the intakes, see Table 7-7. If the effect is statistically valid, it may be caused by participants being sensitized to multi-channel issues by the intake questions. (Most sessions contained intake participants, and the other participants of those sessions scored lower channel coherence. Hence it is less likely that the sessions' outputs or processes have been influenced, than that this is a sensitization effect.)

Table 7-7 Channel coherence averages: session type x intake participation (y/n)

	MuCh-QFD	FE
Intake participation	5.7 (n=6)	4.5 (n=5)
No intake participation	3.9 (n=18)	2.8 (n=20)

For the variable **session timing (morning vs. afternoon)**, we found an interaction effect with session type for the construct customer orientation (Anova, F=5.31; p<0.05), see Appendix I. As illustrated in Figure 7-4, FE session scores are generally lower than MuCh-QFD with regard to customer orientation. However, this effect is decreased for the two FE sessions that were conducted in the morning, causing the interaction effect. This may well be a random effect. We checked the raw data in relation to our own qualitative case analyses and concluded that the participants of the two FE morning sessions have been relatively optimistic in their scoring (especially on items 1a and 1b), in relation to the events as we observed them. Without this 'optimism', the interaction effect would not have occurred. Thus, although the interaction effect reflects a pattern we recognize in the data, the internal validity of this effect is limited: with more cases and observations we expect this effect to disappear. Regarding the main experimental effects we measured (section 7.2.2) the impact of this interaction effect has been limited. We concluded that MuCh-QFD sessions are more customer-oriented than FE sessions. Although without the interaction effect, the difference across both session types would become even more significant, this would not alter our conclusions.

Table 7-8 Customer orientation averages: session type x timing (morning vs. afternoon)

	MuCh-QFD	FE
Morning session	4.7 (n=16)	3.7 (n=9)
Afternoon session	4.9 (n=8)	2.3 (n=16)

Finally, we discuss the relationships between **intention to use results** [question 7] and our constructs. Interestingly, this was one of the few variables where there was a significant difference between the FE and MuCh-QFD groups. Anova generated no interaction effects, but several linear effects: for channel synergy, marketing strategy, progression and focus. Because of the differences between FE and MuCh-QFD groups, we further analyzed these groups separately. Pearson correlation results were different for both groups. For FE, two constructs correlated significantly with intention to use: marketing strategy (r=0.40; n=25; p=0.05) and focus (r=0.65; n=25; p<0.001), see also Table 7-9.

Table 7-9 FE correlations of intention to use results with channel synergy, marketing strategy, progression and focus (n = 25)

		Intention to use results
Channel synergy	Pearson Correlation	0.37
	Sig. (2-tailed)	0.071
Marketing strategy	Pearson Correlation	**0.40(*)**
	Sig. (2-tailed)	0.050
Progression	Pearson Correlation	0.39
	Sig. (2-tailed)	0.056
Focus	Pearson Correlation	**0.65(**)**
	Sig. (2-tailed)	0.000

** Correlation is significant at the 0.01 level (2-tailed).
* Correlation is significant at the 0.05 level (2-tailed).

With regard to MuCh-QFD sessions, four constructs correlated significantly with intention to use: channel coherence (r=0.61; n=24; p<0.01), marketing strategy (r=0.42; n=24; p<0.05), progression (r=0.62; n=24; p=0.001) and focus (r=0.49; n=24; p<0.05), see also Table 7-10.

These correlation results have to be interpreted with care. First of all, because intention to use results is not just an indicator of usefulness, it also expresses the degree of involvement of participants in relation to the e-service of their session. Even though we did not measure this, we know from participant background information and interview results that the intention to use session results varied across participants prior to the sessions. Then the question is: what is the causal meaning of correlation in this context? From our experimental design we cannot conclude what is cause and what is effect. Has higher performance on specific requirements increased the intention to use session results? Or has higher intention to use results generated higher scores on our constructs? Regarding the latter, it is plausible that there is a sensitization effect, analogous to the effect described above for the 'participation in intake' variable. An intention to use session results makes participants more attentive of channel synergy and marketing strategy discussions, since both types of topics involve the evaluation of alternative options for e-service implementation. Implementation options are most relevant for participants who intend to proceed with the e-service idea. On the other hand, for all four constructs (channel synergy, marketing strategy, progression and focus) causality may well work the other way around as well: intention to use session results increases when performance on the constructs increases. If intention to use results is taken as an indicator for session usefulness, then the correlations indicate that FE sessions are judged useful because of focus and, to a lesser extent, marketing strategy. And MuCh-QFD sessions are considered useful because of channel synergy and progress, and to a lesser extent because of marketing strategy and focus. However, we must be careful with these causal conclusions, due to the possible interference caused by the sensitization effects we described.

Table 7-10 MuCh- QFD correlations of intention to use results with channel synergy, marketing strategy, progression and focus (n = 24)

		Intention to use results?
Channel synergy	Pearson Correlation	0.61(**)
	Sig. (2-tailed)	0.002
Marketing strategy	Pearson Correlation	0.42(*)
	Sig. (2-tailed)	0.041
Progression	Pearson Correlation	0.62(**)
	Sig. (2-tailed)	0.001
Focus	Pearson Correlation	0.49(*)
	Sig. (2-tailed)	0.014

** Correlation is significant at the 0.01 level (2-tailed).
* Correlation is significant at the 0.05 level (2-tailed).

In conclusion, our main effects (between session type and requirements customer orientation to stakeholder communication) are not challenged by disturbing factors. Most variables we researched, like for example industry background, firm size, participant expectations, years of working experience, etc, do not appear to have a significant influence on how constructs have been scored. Although this finding must be interpreted with some care (due to the previously mentioned limitations of our analyses of variance), our field experiment appears to be relatively robust. At a statistical level, we found three interaction effects. The two interaction effects for channel synergy appear to be random statistical effects: we do not understand them on the basis of our other observations nor on theoretical grounds. The interaction effect between session type and session timing (morning versus afternoon) was likely to disappear with more observations and thus lacked internal validity. For all three interaction effects we checked whether they had an impact on our conclusions regarding the experimental main effect (7.2.2), which was not the case. Again, the robustness of our findings is confirmed.

Finally, we found four variables to have significant linear relationships with our construct scores (independently of the experimental condition MuCh-QFD versus FE). Regarding stakeholder perspective, our findings are likely caused by sensitization and are in line with common sense: operations/IT professionals mostly ignore competitive positioning details (which is reflected in lower than average scores), whereas the marketing perspective is highly focused on these items (reflected in higher scores). A second linear effect was between satisfaction and focus of the e-service definition process. The direction of this causal relation was determined in combination with our observations: participants become more satisfied when they experience more focus. A third linear effect we found was that participation during the intake increases sensitivity to channel coherence (we inferred the direction of causality on logical grounds). The fourth variable for which we found linear effects was 'intention to use session results'. For FE sessions, two constructs correlated significantly with intention to use: marketing strategy and focus. And for MuCh-QFD sessions, four constructs correlated significantly with intention to use: channel coherence, marketing strategy, progression and focus. This might indicate that FE sessions are judged useful because of focus and, to a lesser extent, marketing strategy. And that MuCh-QFD sessions are judged useful because of channel synergy and progression, and to a lesser extent because of marketing strategy and focus. But there remains uncertainty regarding the direction of causality, since several of these relations might also be caused by sensitization. For example, an intention to use session results makes participants more attentive of channel synergy and marketing strategy discussions, since both types of topics involve the evaluation of alternative options for e-service implementation. And implementation options are most relevant for participants which intend to proceed with the e-service idea. Regarding this fourth variable, intention to use results, our field experiment does not permit definitive conclusions.

When reflecting on all four variables showing linear relations, we can say that our construct scores show some sensitivity to the interests of participants. But this sensitivity is limited enough to not have distorted our main experimental findings of section 7.2.2.

7.3 Overall findings and conclusion

In 7.2 we discussed the answers to the hypotheses for our experimental findings. This was based on rigorous testing, via a research design to test the main effects we expected in response to specific MuCh-QFD design tasks. In this section we first address several more general findings (section 7.3.1). Next we present our main conclusions in section 7.3.2.

7.3.1 Overall findings

An unexpected finding from our research is that we found that is it difficult for business participants to gauge the strong and weak points of a design support approach if they do not have other approaches as a reference point. For MuCh-

QFD participants 'the logical approach' was: starting from customer needs and for FE participants it starting from e-service objectives. This is indicated by the agenda item usefulness scores of participants in section 7.1.5: in FE as well as MuCh-QFD sessions the first agenda item is considered most useful 'since it is the basis' in the perception of participants. Participant preferences are likely influenced by the introduction in each workshop, where the approach that will be followed is briefly explained. Having said that, it came as a surprise to us how easily either approach is automatically accepted as 'the logical approach' by the majority of the participants.

Another point that we want to highlight is that many questions of positioning, scope, customer needs and service priorities that are addressed in these early service definition phases are at least as much marketing questions as they are design issues. Nevertheless, our participants are positive about the usefulness of deploying design methods like MuCh-QFD. They appreciate the fact that discussions do not last very long and are followed by group votes or other methods of generating preliminary design choices. Based on those choices a new 'relative certainty' is created from which to proceed. Participants liked creating those preliminary certainties, even if those certainties were only relatively stable and could be changed afterwards. Hence, the paradox is that although we mainly leaned on design methods in terms of structure and method, we tackled many marketing issues in our sessions. How complete those marketing discussions were, is still an open question and falls outside the scope of our research. However, it is interesting to see how much our design methods contributed to solving issues in the area of service marketing and management.

We also found that marketing strategy issues are addressed more in FE than in MuCh-QFD sessions, even though FE originates more from technical engineering than QFD does. In our sessions it became apparent that QFD is a relatively strict approach (in comparison to FE): it provides much guidance on how to connect customer priorities, service functions and performance levels, competing offers, and in our case multi-channel needs and functions, but the QFD format itself does not stimulate digression into more abstract marketing strategy discussions. It is also interesting to note the difference with our pre-test results (5.4.2), which showed that MuCh-QFD sessions contributed significantly more to the original construct of 'competitive positioning' than FE sessions. We attribute this to the fact that pre-test participants were students and that the case assignment was an artificial problem for them. Hence, there was less need for those students than for business participants to address competitive positioning issues, and FE sessions with students showed fewer competitive positioning discussions. Overall, this difference between pre-test and field experiment results confirms the prudence of choosing 'real world' problems and participants to test MuCh-QFD performance with.

The output of both sessions was considered relatively complete (participants were unanimously positive about the amount of progress in relation to the time spent), but not detailed enough to be a final e-service design. Elements that participants said were missing were, for example, address how to implement the e-service in the back-office, refine the business case, define Internet pages and navigation structure, and further investigate competitor e-services and positioning. This confirms our overview of design tasks in Figure 4-5, which shows that the initiation phase should be followed by a development phase which contains many of the elements mentioned here, before the total design process is completed. Also, in several sessions (most notably the FE sessions, with explicit site definition in part III), the challenge arose of how to make the e-service simple and easy to use for customers. This was a challenge that was generally left unanswered in our sessions. It is addressed in one of the first tasks following the initiation phase: in the development phase, when the storyboard (including site navigation) and visual prototype are defined (see also our suggestions for further research in the next chapter).

Defining the core solutions on the Internet site in part III of FE does not consistently generate high quality output. The quality depend very much on the presence of a participant who has clear ideas about the e-service (cases A: absence management, C: intermediary portal, E: online billing) or who has a good ability to facilitate and integrate the feasible ideas of others (case B: insurer portal). Other examples of session dependence on the quality of participants are: the 'traveler from the future' in case C, FE session, and the sales person strongly advocating customer priorities in case B, MuCh-QFD session (see section 7.1.1). The more free-format discussions are, the larger the influence of individuals on process as well as output.

As discussed in section 7.1.5, the participants' opinions about the service slogan definition varied between MuCh-QFD and FE participants. We explained that this was (at least partly) due to its relative position in the session. This finding provides empirical support for the intuitively logical idea that the value of a design task depends on its position relative to other design tasks (which makes it less straightforward to determine 'the' value of a design task experimentally). One question is how it translates previous analyses into design consequences, and another question is how directly and integrally it connects to subsequent design decisions.

7.3.2   Conclusion

In this chapter we answered our sixth research question - how MuCh-QFD performs in relation to the design support requirements customer orientation to stakeholder communication. In summary, MuCh-QFD performs relatively well in terms of customer orientation, channel coherence and communication between stakeholder perspectives. Contrary to our expectations, we found that the control group session scored higher in terms of marketing strategy discussions (though not on competitor focus). This can be contributed to FE tasks that addressed objectives (part I of FE) and the question whether the newly defined e-service merited investment or not (at the end of FE session part III). In MuCh-QFD sessions the discussions regarding competitive positioning focus more on explicit comparisons with competitor e-services. Both session types received much praise from participants for the amount of progress and focus that was generated via the structured format that was used, which was enhanced by our use GroupSystems™. The QFD 'service' or 'relationship matrix' was particularly valued for increasing customer focus, supporting investment decisions and communicating the concept as well as design decisions to others after the

155

session. Especially managers were positive about its usefulness. In cases where multi-channel aspects were not prominently present in the initial e-service idea, MuCh-QFD really added value and created several eye-openers regarding channel interdependence. Finally, the intake protocol also contains service definition questions which proved to contribute to customer orientation, cross-channel awareness, competitor awareness, financial attractiveness of the service idea, and creating focus for the e-service definition tasks in the sessions.

What our research has also shown is that when the FE participants are left to themselves to determine objectives, functionalities, online solutions and how to position the e-service, they tend to focus on supplier interests (even despite the multi-disciplinary setting we provided). The only exceptions are cases where there are very specific factors present that enforce customer or multi-channel attention. Since the feedback from participants indicated that FE sessions have a higher process quality (at least in terms of our requirements) than 'regular' process quality in firms, this is an ominous indication for the degree of customer or multi-channel orientation in average e-service definition processes. Hence MuCh-QFD can be expected to have performed even better on our design requirements in comparison to the 'real world' e-service definition processes than in comparison to our control group sessions.

To conclude, our field experiment has shown that our MuCh-QFD method increases design process reliability in comparison to FE and common practices in firms, measured in terms of our design requirements. (The exception is that FE sessions show a higher degree of marketing strategy discussions, but this is not always a sign of strength. For example, in case F: telecom bundle, there was much strategy discussion, but this was mainly due to uncertainties regarding e-service objectives and priorities.) In other words, our conclusion is that design process variability is lower with MuCh-QFD. More specifically, the commonalities that FE and MuCh-QFD share (e.g. intake protocol, using a design method, session with strict agenda and multiple stakeholders) already improve the reliability of the design process. This decreases the chances of overlooking for example customer, supplier or technology issues. And on top of that, MuCh-QFD is more reliable than FE in relation to our design requirements. Looking back at the various cases, we conclude that, although in every case the e-service design challenge that is most apparent will emerge and be addressed in both sessions, the other challenges are only addressed seriously in MuCh-QFD sessions. For example, cases B: insurer portal, and C: intermediary portal, had service ideas that were meant to counter serious market pressures. Those cases also had the most serious discussions about how to make the e-service successful and how to position it in the market. And case E: online billing, was so inherently customer-oriented that this came up in both sessions. However, the complete range of stakeholder interests is addressed only rarely in FE sessions, due to the focus of the participants. Looking back at the findings of chapter 3 regarding our explorative research into 19 cases, where we reported that 'regular' e-service design processes often tend to become one-sided initiatives, we can say that that is confirmed by our experimental findings. What MuCh-QFD adds is balance: it addresses multiple issues in relation to each other, and helps avoid single focus initiatives that have a higher risk of failure because several relevant aspects have not been properly incorporated into service definition choices.

156

8   Conclusion

In theory there is no difference between theory and practice. In practice there definitely is.
(Anonymous)

Case Exhibit 8-1 Design methods only tell part of the story: example of participant influence

Case C, Intermediary Portal, had an FE session participant that could be classified as 'a traveler from the future'. His company had started planning a similar e-service four years ago, and it was commercially operational since two years. He provided many insights into a) what is a clever way to tackle problem XYZ?' and b) what do customers want, buy, and respond best to?' This was rather unique and did not occur in any of the other sessions. His input had a marked impact on the session. For one thing, he strongly advocated customer focus and the imperative to connect all service choices to customer priorities. Due to his input the FE results were relatively customer-oriented and comparable to the MuCh-QFD results (and on some points even more specific).

In section 8.1 we discuss the research objective and main findings of our research. In section 8.2 we discuss the theoretical contributions, followed by the limitations of our research (8.3). Finally, we have several suggestions for research in the areas of design research and design research methodology (8.4).

8.1   Research objective and main findings

Surprisingly enough, service design is among the least studied and understood topics in services marketing, even though service quality is crucial to loyalty and profit (Heskett, Sasser et al., 1997), and service design has been identified as 'perhaps the most crucial factor for quality' (Gummesson, 1993). Furthermore, in recent years research has shown that customers have increasingly become multi-channel shoppers, and find it quite natural to use websites and physical stores as part of the same buying process (Schueler, 2003). This multi-channel domain is the focus of our research. We started our research into 'click and mortar' practices in 1999, when most people still believed that 'the dotcoms were going to conquer the world'. Now that the bubble has burst, click and mortar approaches have become a more general phenomenon. However, businesses still struggle to find the best channel mix.

Our *research objective* was to help organizations design auxiliary Internet services (or e-services). Although e-service design is often considered in isolation of other channels, we have demonstrated that e-services should be aligned with services provided via other channels and not be designed in isolation. In answer to our research question - how to design e-services functioning in a multi-channel context - we developed MuCh-QFD, consisting of an intake and a four hour design session with multiple stakeholders, based on QFD, with multi-channel additions. We compared the performance of the MuCh-QFD session with control group sessions on seven design requirements (customer orientation to stakeholder communication). The control groups used a 'Fundamental Engineering' (FE) design approach.

Our research has different types of findings. In section 8.1.1 we discuss findings in response to the empirical question: 'Do MuCh-QFD sessions perform better on our design requirements than FE sessions, and can we expect MuCh-QFD and FE sessions to have performed better than 'regular' service definition processes in firms?' These findings have implications for management, which we discuss in section 8.1.2. In section 8.1.3 we summarize the main findings in response to the design knowledge question: 'Which elements of MuCh-QFD (and FE) are effective in relation to our design requirements and why are they effective?' And in section 8.1.4 we discuss the answers to our research questions introduced in section 1.2. Those answers summarize our findings from chapters 2 to 7.

8.1.1 Main research findings

The main causal question relating to our field experiment is: 'Do MuCh-QFD sessions perform better on our design requirements than FE sessions, and can we expect MuCh-QFD and FE sessions to have performed better than 'regular' service definition processes in firms?' In answer to this question, our research shows that:

- The FE session process is judged by participants to be of higher quality (in terms of progress, focus, usefulness of results and satisfaction in comparison to previous service definition experiences) than 'regular' service definition processes.
- We have shown experimentally that when groups are left to themselves (even the multi-disciplinary groups of FE sessions) to determine objectives, functionalities, online solutions and how to position an auxiliary e-service, they tend to focus on supplier interests (except when the explicit goal of an e-service was to improve support for customers or other channels). Thus, although FE sessions have higher than 'regular' process quality and actively involve multiple stakeholders, customer-related and multi-channel issues are more often than not left untouched. In connection with the previous finding this implies that 'real world' e-service processes are likely to ignore these issues as well, which is confirmed by our explorative research findings across 19 'click and mortar' cases. This has serious management implications, which we discuss in section 8.1.2.
- MuCh-QFD sessions perform significantly better than FE sessions on customer orientation, channel coherence and communication between stakeholder perspectives. And MuCh-QFD participants explicitly state that

these are valuable contributions. Moreover, they appreciate the way the MuCh-QFD service matrix helps communicate to others the choices on the basis of the various perspectives after the session is completed. This shows that the additional attention to customer and multi-channel considerations in MuCh-QFD sessions is not 'artificial', but that it has business value.

8.1.2 Implications for management

Our findings have several important implications for management. Firstly, if customer and multi-channel issues are insufficiently addressed in FE sessions (under high quality process conditions) it is very likely that the same will happen in 'the real world'. Secondly, this means that in practice, even though many people pay lip service to customer orientation, it is often no more than just that: there tends to be a bias towards the supplier's point of view and optimizing services based on supplier priorities rather than customer or channel partner priorities. The consequence of this, as illustrated in chapter 3, is often a waste of efforts and resources due to the development of unsuccessful e-services: the e-services are not adopted by customers, they are not supported by channel partners, or they erode profit margins for the suppliers as a result of the fact that not enough thought has been put into how the service ought to be positioned.

From these findings we conclude that our MuCh-QFD method, which is fast, has low entry barriers (in terms of participant training or resources required) and performs well in relation to customer and channel issues, adds significant value to current practices of new e-service definition.

8.1.3 Main design knowledge findings

The main question addressed in this section is: 'Which elements of MuCh-QFD (and FE) are effective in relation to our design requirements and why are they effective?' This relates to the main 'design method' lessons that can be learned from our research. According to van Aken (2005), design knowledge can be formulated in the form of 'technological rules' or 'solution concepts' that follow a basic logic of "if you want to achieve Y in situation Z, then perform action X". Moreover, the rules must be 'grounded' in a wider body of knowledge that explains why the application of the rule gives the desired result, in order to transcend the level of mere 'instrumentalism'. Hence, design knowledge should be given with 'thick descriptions' (Geertz, 1973), discussing (dis)advantages of solution concepts under various conditions. This facilitates translation of general design principles to specific contexts (Aken, 2005).

We start by briefly reviewing our design method and its genesis. In chapter 3 we concluded that decision making during the initiation phase of new e-services needed improvement (in terms of the design requirements we identified). We evaluated several existing methods in relation to those requirements and concluded that a modified version of the QFD (Quality Function Deployment) approach would likely perform best (see section 3.3 for an evaluation of strong and weak points of different methods). Via modifications to QFD, two concerns were tackled: lack of speed and lacking incorporation of multi-channel issues.

Hence, we developed a session format that only requires four hours time investment from participants. And multi-channel considerations were incorporated in part III of our MuCh-QFD session. For a full description of our MuCh-QFD method, the reader is referred to section 4.2. An overview of the expected contributions of MuCh-QFD sessions in relation to the design requirements is presented in Table 5-3. For a description of how our MuCh-QFD method was pre-tested in three rounds and which considerations guided the evolution to its final format, see section 5.4.

Looking back at our pre-test experiences, these are several 'technological rule' or design method lessons to be learned:
- When defining a multi-channel service mix, focus and progression in the design process benefit from defining the e-services first and then adding multi-channel services, instead of trying to define all channel services in parallel. (Our first pre-test showed that the latter approach requires relatively much 'mental processing power' of participants and creates divergent attempts of participants to be exhaustive in the number of options that can be identified, instead of stimulating prioritization and decision making.)
- When testing design methods with groups, make sure to have strict process control and precise task definitions. This increases repeatability, which a) increases design process reliability, b) enables reliable testing of the method because it is a stable method, and c) makes the method 'teachable': others can use the method and obtain similar results.
- When creating service definitions in the initiation phase of a design process, avoid detailed cost discussions. A high level business case is useful to understand (and possibly optimize) the profit potential of the service, but detailing the business case should wait until at least the development phase. In the initiation phase too many design details are still unknown for accurate business case detailing.

These lessons are the product of theory generation during our pre-tests and have only been field-tested in a qualitative way during our field experiment. They were not part of our research design as defined in chapter 6. Hence, there is room for more rigorous experimental testing of these 'technological rules.'

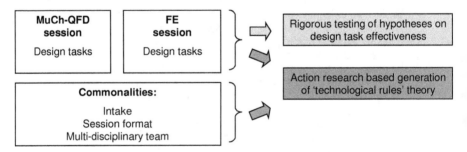

Figure 8-1 Summary of the research design that has been used

In Figure 8-1 we summarize the research design for our field experiment. As illustrated in the top half of Figure 8-1 the research design we used allows for rigorous testing of hypotheses on effectiveness of design tasks that are part of the session agendas. However, other lessons we draw from our experiences are better described as action research based theory generation. MuCh-QFD and FE groups share many design method commonalities as illustrated in the bottom half of Figure 8-1. They use the same intake protocol and outputs, and the session formats are also similar (in terms of type and number of participants, using the GroupSystems™ decision support system, or structure and length of the sessions). Hence, the research design does not allow hypothesis testing with regard to those commonalities.

Based on our field experiment, the main design method findings are:
- The expected contributions of MuCh-QFD sessions in relation to the design requirements as presented in Table 5-3 are largely confirmed experimentally. There are two exceptions. Firstly, whereas channel coherence was stimulated, channel synergy was not (at least not significantly more than in FE sessions). MuCh-QFD participants indicate a possible reason: they thought it to be too early in the design process to discuss channel synergy details extensively. Secondly, competitive positioning was not stimulated as a whole, but had to be split into two different constructs: 4a, competitor focus, and 4b, marketing strategy. When considering our qualitative data, it appears that MuCh-QFD sessions stimulate competitor focus by explicitly comparing competing e-services, but the difference with FE sessions was not large enough to be statistically significant. On the other hand, FE did stimulate marketing strategy discussions significantly more than MuCh-QFD. This can be attributed to two FE session agenda parts (see also Figure 7-5): part I when objectives are discussed and especially part III when discussing whether the e-service is 'investment ready' or not. A related benefit of the latter agenda item is that other managerial requirements for success also emerge from the discussion. This is useful input for the next design phase: development. Based on these findings we would recommend inclusion of this agenda item in a future version of the MuCh-QFD agenda at the start of agenda part four.
- Our field experiment has confirmed that the main reason for usefulness of the MuCh-QFD session agenda is: starting from customer priorities, explicitly connecting them to service functionality priorities, incorporating multi-channel customer needs and functions, connecting stakeholder perspectives in the relationship matrix and (which is formally outside the scope of our field experiment, but was indicated by multiple participants:) aiding the communication of design decisions to others after the session with the use of the relationship matrix.
- Regarding the FE session we have experimentally shown that business participants tend to focus on supplier needs instead of customer needs when moving from objectives, to functions, to online solutions. This is probably partly caused by the fact that customer priorities do generally not represent the majority in the objectives list that is created in part I of FE sessions. Another reason appears to be that professionals are generally more inclined to focus on supplier than on customer advantages, unless 'forced' to do otherwise.
- We have made plausible that the usefulness of the session format is related to the limited time investment required from participants, to promoting shared

understanding and commitment between different stakeholders, and to the fact that attention is drawn to several relevant issues in one subsequent process that leads to priorities and focused outputs on each of the design tasks. (These findings are confirmed by feedback from participants from MuCh-QFD as well as FE sessions. Still, these claims are the product of theory generation. Our experimental design was unsuited to confirm or falsify them.)

- The main reason for usefulness of the intake protocol appears to be that it provides focus (which product-market combination, which buying cycle phase, start from a specific e-service instead of a broad multi-channel strategy) and that it aids in making preliminary strategic positioning choices. The strategic positioning choices are presented at the start of a MuCh-QFD session and can be questioned and/or corrected by participants when necessary. (This are also claims based on theory generation. Our research design was unsuited to confirm or falsify these specific claims.)

- An overall rule that emerges from the experiences across pre-tests and field experiment sessions is: in group sessions, do not try to solve more than 2 design challenges simultaneously, but use elimination. Solve issues one by one in a serial approach, instead of a parallel one. If there are dependencies, make an assumption and proceed. Wait until later to check for overall coherence and what the consequences would have been of having made other assumptions. The main reason for this serial approach is to maintain group focus and progression: it appears as though groups have even more difficulty than individuals to address multiple challenges at once. (This claim is an invitation to further research.)

Above, we summarized the main design method lessons. However, as stated in the beginning of this section, design knowledge needs 'thick' descriptions to illustrate the effects of design methods under various conditions. In other words, more nuance and detail are needed. This can be found in sections 7.1 and 7.3; they most prominently provide 'thick' design knowledge. For example (section 7.1.1): "in the FE session for case A, defining the service slogan was relatively useful in comparison to other cases: in light of the relative complexity of the service it helped to formulate the core intention of the service explicitly." Along these 'thick' descriptions, section 7.2 provides a confrontation between the results from our hypothesis testing and the theory generation results from our qualitative analyses. Furthermore, in section 8.2 we summarize the contributions and limitations of MuCh-QFD for each of the seven design requirements.

8.1.4   Answers to research questions

In this section we discuss the answers to our research questions introduced in section 1.2. Those answers summarize our findings from chapters 2 to 7.

Our overall *research question* was:

How to design e-services in a multi-channel context?

This question was broken down into several sub-questions. These questions are:

1. What determines the competitive value of a channel within a multi-channel mix, and where can opportunities for the Internet be expected? *(This question is addressed in chapter 2.)*
2. How can design methods aid the process of designing new e-services that have to function in a multi-channel context, and what are the requirements for such a design method? *(This question is addressed in chapter 3.)*
3. How can we develop an e-service definition method that meets the requirements mentioned in the previous question? *(This question is addressed in chapter 4.)*
4. Which e-service definition method can we use as a control group condition? *(This question is addressed in chapter 5.)*
5. How can we evaluate the performance of our (experimental and control group) e-service definition methods with regard to the design support requirements? *(This question is addressed in chapter 6.)*
6. How do the e-service definition methods perform with regard to the design support requirements? *(This question is addressed in chapter 7.)*

The answer to our first research question, regarding the competitive value of channels, has two main points. Firstly, different channels contribute to a channel mix in different ways. Their contributions also vary per phase and 'service element' (like 'need identification' or 'transferring possession of goods' or 'after sales usage advice') in the buying cycle. When the various service elements are evaluated, different (dis)advantages become clear for each of the channels, which makes their value-adding potential explicit. Moreover, what suppliers perceive as the advantage of a channel for a supplier is not always identical to what customers perceive them to be. The points of view of suppliers and well as customers have to be considered when designing new e-services.

Secondly, our explorative case study results show that we should always consider the local context of a case when defining new service opportunities for the Internet. On the one hand, the added value of a new channel should be evaluated explicitly relative to the existing channel mix. For example, if one lives next door to a supermarket, online ordering and home delivery may not always add much value. However, if the desired goods are not available nearby, it may be quite attractive to use online ordering plus home delivery. On the other hand, the value for a customer depends on customer context and preferences, which also vary per case, and even across groups within a case.

Thus, identifying Internet service value is a generic task only to a certain extent. There is always a serious design task that involves making a case-specific analysis of the 'design problem' and then moving towards developing suitable solutions.

The answer to our second research question - which method could support this design process and what the requirements for such a method are - is that a) a lean version of QFD, with multi-channel additions can be expected to aid the process of designing new e-services, and that b) requirements for design support are: 1) customer-oriented design, 2) channel coherence, 3) channel synergy, 4) competitive positioning, 5) speed, 6) focused design process, 7) stakeholder communication, and 8) communication of service concept coherence during implementation.

The answer to our third research question - what would be a suitable method - is that such a design support method can be developed as a structured intake plus a four hour session with multiple stakeholders using GroupSystems™, based on a QFD approach with some multi-channel extensions (MuCh-QFD). From the QFD 'House of Quality' it is mainly the initial 'rooms' that we use as a basis (rooms, 1 to 4, plus 6). Our multi-channel extensions are situated in rooms 1 to 3 of the House of Quality.

The answer to our fourth research question - which control group method should we use - is that the control group design process should also adhere to a session format (instead of, for example, the 'regular' process participating firms usually follow). We also concluded that the approach used in such a session should be an alternative design method (instead of, for example, using no predefined method). More specifically, we found 'Fundamental Engineering' to provide a suitable basis for our control group sessions.

In answer to our fifth research question - how to evaluate MuCh-QFD sessions - we used multi-method data collection in an experimental design very close to 'static group comparison', using Fundamental Engineering (FE) sessions as control conditions. To test the MuCh-QFD sessions we used six cases with business participants who had an e-service innovation idea with multi-channel implications. The design team of each case was split into two groups of four to five participants. For each case one group attended the MuCh-QFD session and the other group attended the FE session. We used observers and participant questionnaires to collect data, we interviewed participants after the sessions, and we were provided feedback by the participants during a debriefing several weeks after a design session. The measurement instruments we used for the observers and participant questionnaires were based on our design requirements. The instruments as well as the design sessions were pre-tested before our experimental phase with the six cases. Afterwards, we evaluated the validity of the measurement instrument on the basis of our experimental data.

With regard to our sixth research question - how does MuCh-QFD perform in relation to our design support requirements - MuCh-QFD generally speaking performs relatively well in terms of customer orientation, channel coherence and communication between the various stakeholder perspectives. Surprisingly enough, the control group session FE scored higher on marketing strategy discussions (though not on competitor focus). Participants expressed their appreciation for both session types for the amount of progression and focus that was generated via the very structured format that was used, which was enhanced by using GroupSystems™. The 'service' or 'relationship matrix' we adopted from QFD was particularly valued for increasing customer focus and concept manageability. Especially managers were positive about its usefulness. In the cases where multi-channel aspects were less prominently present in the initial e-service idea, MuCh-QFD really added value and offered participants new insights into the interdependence between channels. Finally, the intake protocol used in preparation of the sessions also contains service definition questions which contributed to customer orientation, cross-channel awareness, competitor awareness, financial attractiveness of the service idea, and created focus as far as the e-service definition tasks in the sessions were concerned: a MuCh-QFD session without the structured intake would lose part of its value.

8.2   Theoretical contributions

In this section we start by summarizing the contributions and limitations of MuCh-QFD per design requirement (customer orientation to stakeholder communication). This provides further detail to some of the 'technological rules' that express design knowledge lessons in section 8.1.3. Next, we discuss the theoretical contributions on a more overall level.

With regard to our design requirements, we have demonstrated the effectiveness of a customer priorities list in combination with the e-service matrix (which is comparable to HoQ room 3) in stimulating customer orientation, which is our first design requirement. There are methods from, for example, the field of User Centered Design (Vredenburg, Isensee and Righi, 2001) of user task modeling (Beyer and Holtzblatt, 1998) that delve deeper into user contexts, but what MuCh-QFD adds is the explicit linking of customer priorities to functionality priorities and competitor comparisons. Especially managers valued the e-service matrix as a basis for investment decisions and to communicate the e-service concept to others. Secondly, our research has shown that explicitly adding multi-channel related customer needs and functions next to Internet services is useful for click and mortar situations. One could say that QFD is originally bipolar with regard to the stakeholder interests it incorporates: customer priorities and competitive success for the supplier. Our multi-channel coherence extension (namely, not just the supplier and customers, but also channel partners) is a first step towards the integration of more than two stakeholder perspectives into design choices. As we discuss below, this is a promising area for future research. Next, in relation to our third design requirement our results show that extensive channel synergy discussions are not useful as a group exercise during the initiation phase (and during our pre-tests we found similar results with regard to detailed discussions on financial implications). Participants feel that they have few useful things to say on these topics, as long as the e-service definition, development efforts and implementation impact are not detailed enough. At least not in addition to the high level (financial) impact estimates that are made during the intake. Fourthly, MuCh-QFD addresses competitor comparisons in a relatively functional way (HoQ room 4). Although this is useful, it not sufficient to cover all the relevant competitive positioning challenges of the initiation phase. Other important questions that need to be answered are: How does the e-service contribute to our objectives and our overall marketing strategy? Is the e-service attractive enough for other business partners that are needed for e-service success? Or does it cannibalize on existing channel services? And how will we make money with this e-service? In other words: although comparing an e-service with what competitors have to offer is useful from a strategic marketing perspective, it only covers part of it (Kotler, 1999). Several of these questions are addressed during the intake, but it is prudent to come back to them after completing the MuCh-QFD session, during the development phase tasks of Figure 8-2. At that point information regarding customer value, competitive options, feasibility and implementation efforts has become more precise, which enables the necessary strategic marketing decisions.

With regard to the design process, our results show that the speed and focus of MuCh-QFD are highly valued by participants. Hence, we have shown that using a four hour session format, supported by GroupSystems™, with four different stakeholder perspectives present, and in connection to an intake that addresses several services marketing questions (like target segment, service objectives, service blueprinting, financial attractiveness) does not deteriorate the business value of QFD, but rather makes some of the QFD and services marketing basics available for participants at limited costs. Also, the integration of different stakeholder perspectives in this early phase was appreciated. In literature it has been said that QFD has limited value if it is used as 'just an exercise to fill in the numbers' and it was more important to embrace the overall QFD philosophy (Clausing, 1994; Ramaswamy, 1996). We agree with this opinion, but interestingly enough our results show that even 'just filling in the numbers' and discussing those results during a four hour session is valuable to participants, especially during the service initiation phase and especially when multiple stakeholders are present (to avoid one-sided views). Furthermore, the impact of the 'number–filling' exercise has been that several participants expressed an interest in the QFD method, and that part of our MuCh-QFD philosophy (namely to start from customer priorities, explicitly link them to functional priorities, and to integrate multi-channel needs and functions into the e-service definition choices) became apparent and attractive to several participants. They expressed an appreciation for the possibilities that MuCh-QFD offered them to evaluate investment choices and communicate the 'what' and 'why' of e-service definition choices to others. Overall, it is interesting to see that QFD added so much value during the e-service initiation phase, when it was transformed into our MuCh-QFD format. By contrast, QFD originated from a need to integrate the entire development process by linking the development and implementation phases to initiation phase decisions. Although we focused exclusively on the initiation phase, MuCh-QFD proved useful, according to the participants as well as our own measurements and observations. In chapter 4 we argued in favor of using a session format for the e-service definition process in the initiation phase, to connect to the usual practices of business participants. Our process performance findings and feedback from participants indicate that we made the right decision.

Reflecting on our contributions to theory at a more abstract level, there are several things we would like to mention. Firstly, we made the tasks for e-service definition in the initiation phase very explicit, (see also Figure 8-2), and tested them experimentally (which also showed us that group exercises on channel synergy or financial feasibility have limited value in the initiation phase: they add more value in the development phase, see also our discussion on further research below). Secondly, we developed a method that also appears to be more broadly applicable for defining auxiliary services via other channels and technologies like, for example, mobile services, speech recognition services in call centers, or even new 'personal contact' services that are added to the channel mix (as illustrated by ING which started ING DIRECT Cafés to be 'closer' to customers and complement its online presence). What is generic is that the value of a new auxiliary service is evaluated in relation to the existing service mix and competing options (in terms of functionalities and contributions to customer priorities). For example, in case D: offer registration, we compared the e-service against a paper-based procedure as one of the competing offers. Thirdly, we have shown experimentally that professionals responsible for defining new e-services tend to focus on supplier priorities instead of customer priorities. (There is an interesting analogy with psychology findings: what people say they would do in situation X, is often not what they actually turn out to do. Likewise, paying lip service to customer orientation is widespread in service development, but reality tells a different story.) Thus, a method that 'enforces' customer orientation is valuable. Hence, as a fourth point, we have also shown the importance of developing methods that support organizations in defining e-services. This confirms the three central elements from which we started our research (see section 1.4), which culminated in the third element: the design of (channel) services can and should be supported by well-structured and well-tested design approaches. (Preferably these approaches should be fast, and with low entry barriers, see also chapters 3 and 7.) And since services imply 'process consumption' for customers, service design methods may need even more customer orientation than product design methods.

8.3   Limitations

There are several limitations to our research. To start with the limitations in relation to our measurements: Firstly, we used process measurements instead of output measurements. Hence, if we measure that MuCh-QFD significantly outperforms FE in terms of customer orientation, for example, this means that the design process has been more customer-oriented, but it does not necessarily mean that the service definition output of the design session is more customer-oriented. We circumvented this to some extent by performing our own subjective output comparisons. The good news is that we found that there is a strong correlation between process performance on requirements and output performance, but this finding is not as rigorous as it would have been if we had used a more objectified measurement instrument. Secondly, our measurement instruments were not based on validated scales from literature. To our knowledge, these do not exist, which is why we developed measurement tools for our constructs. Thirdly, several requirements had to be interpreted slightly differently to become specific and sufficiently meaningful. Thus, we translated speed into a degree of progress and inter-stakeholder communication focused on the degree of team agreement that was reached on key perspectives like customer, channels and competitive position. Hence, our measurements consciously contained a certain bias towards our experiment (see also section 6.1.3). Fourthly, our measurement instrument validation has a limited statistical basis. Due to the nonparametric nature of our items we could not perform a factor analysis to analyze construct validity (see also section 6.4.3). We did perform a reliability check that was empirical in nature: to what extent do qualitative case analyses, the questionnaire measurements and the observer scores show similar results? The advantage of our multi-method approach is that we were able to perform such a consistency test. Finally, at a very practical level, we asked participants about their intention to use the results after the session, although we should also have asked this prior to the sessions. Since we failed to do this, and since during our research we observed that several participants were already prior to the sessions determined to use the results, we were unable to draw conclusions as to how much our sessions contributed to the participants' intention to use the results (an indicator for session usefulness).

A second group of limitations has to do with the experimental design. Firstly, participants experienced only one method, so the option to let them explicitly evaluate differences between both methods did not exist. This is one of the disadvantages of using the 'static group comparison' experimental design. Another approach could have been to let each group design two different e-services, using one methods after another (group 1 first uses FE for service A and then MuCh-QFD for service B, and group 2 first uses MuCh-QFD for service A and then FE for service B). In future research this may be a useful experimental design. Secondly, there were many similarities between the two test groups (for example service definition structuring via the intake, use of the GroupSystems™ decision support system, and multi-disciplinary teams were similar for both groups). Nevertheless, in our qualitative analyses they were confirmed to be important elements of the effectiveness of our MuCh-QFD method. Hence we believe that the differences between our two methods in some respects are relative small in comparison to differences with other service definition approaches (e.g. like giving a product manager in a large corporate the assignment to fill in a business plan template for a new service). In conclusion, our experimental design tested the effectiveness of only part of our MuCh-QFD method. Hence conclusions on effectiveness of the design support elements that were similar across both case groups had to be drawn on the basis of a less rigorous case analysis approach and evaluations of participants in comparison to their previous experiences.

And maybe some readers feel that a third type of limitations relates to the 'straw man' design method (FE) we used for our control groups: "If we consciously choose a control group method that is 'agnostic' in terms of customer needs, supplier needs and channels, then is it any wonder that MuCh-QFD scores higher?" On the one hand we would like to reply on a methodological level. Since the design field is relatively theory poor in terms of rigorously tested design methods, we aimed to approximate a real 'placebo' (not containing any form of 'medicine' for the 'ailment's we were focusing on) as control condition. Simply because we could not reliably predict which design tasks would contribute what on the basis of robust, empirically tested theories. And on the other hand we would like to answer to the validity concerns: "Your MuCh-QFD findings are not valid, because the reference point was a 'straw man' FE session, and your measurements were focused on requirements that also formed the basis for developing MuCh-QFD." There are a few things that speak in favor of the validity of our results. Firstly, the FE approach we used as a basis for our control sessions is a real method, and participants were generally positive about the control sessions. In other words: though agnostic with regard to customers etc, the FE method was valuable to participants. Secondly, FE sessions did not prohibit customer orientation (or focus on any other requirement). Participants were free to choose which type of objectives, functions and online solutions to discuss. Thus, we did not artificially make them less focused on customers etc than they normally would have been. Thirdly, not all our hypotheses were confirmed. (For example, MuCh-QFD sessions did not stimulate channel synergy discussions as much as suspected, which was also visible in our measurement scores.) Fourthly, our measurements led to a double surprise in the sense that construct 4 had to be split up and that the resulting construct 4b, marketing strategy, scored higher for FE sessions than for MuCh-QFD sessions. This was totally counter to our expectations previous to the field experiment, which also confirms the validity of our research design.

8.4　Recommendations for further research

In this section we distinguish two different areas for further research. First we address the area of design theory. And then we discuss research opportunities in the area of design theory methodology.

With regard to design theory, we found that there are several important questions in the area of: which design tasks should be performed when, in order to support an efficient and effective design process? In our research we found for example that detailed discussions on e-service costs or channel synergies are not very fruitful in the initiation phase (see Figure 8-2). Next, it is also interesting to ask 'how' certain design tasks should be supported, and which design methods are most suitable for which types of design challenges. See below for examples of these types of questions.

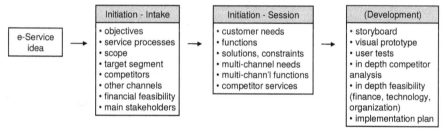

Figure 8-2 Initiation (MuCh-QFD) versus development phase (post-MuCh-QFD)

As stated before, MuCh-QFD focuses on the initiation phase. As an extension of the underlying research, a follow-up session has already been developed and tested. The follow-up session is aimed at the first steps of the development phase: the storyboard and visual prototype of Figure 8-2. The session takes MuCh-QFD output as a basis. It then moves from functional hierarchy and task analysis towards a storyboard with the main Internet pages. In this process personas (prototypical customers from the target segment) are used to help participants in their efforts to realize a customer-oriented design. Subsequently, participants explicitly define links with multi-channel services and back-office connections for each Internet page. One of the main outputs is a visual prototype of the e-service in its multi-channel and organizational context. Whereas MuCh-QFD focuses on generating a common conception of the 'what' of an e-service with multiple stakeholders, this second session helps making the 'how' of the e-service concrete. The resulting visual prototype is valuable for concept testing with customers, for communicating the e-service concept to others, and for providing a detailed e-service definition to facilitate the marketing strategy choices we mentioned in section 7.3.1 on the basis of competitor analysis, feasibility studies and an assessment of the implementation impact.

A second extension we also already embarked upon is a deepening of stakeholder perspective analyses and the question as to how to integrally connect the results of these analyses to MuCh-QFD design tasks. This is triggered by a potential weakness of MuCh-QFD: it focuses predominantly on customer priorities. Case D illustrated that it is less focused on advantages for intermediaries (which would stimulate service adoption by intermediaries). It is an interesting question to check how the method can be adapted to include intermediary priorities as well. This may be possible by adding intermediary needs in the first 'needs' column of the service matrix. In this way, functionality priorities for intermediaries can also be determined. In general, MuCh-QFD provides such a useful basis for linking different perspectives to design choices that it is interesting to see to what extent this can be extended (maybe priorities of more parties can be included in an equally explicit way).

A third promising option for future research has to do with our good fortune to have 'a traveler from the future' participate in the FE session for case C: intermediary portal. It appears that for any new service it is very valuable to obtain contributions from someone with relevant experience. Such a person can provide many insights into questions like a) what is a clever way to tackle problem XYZ?' and b) what do customers want, buy, and respond best to?' It appears that this could reduce the 'time to market success' for new services with months to years. However, we found that the group process did not show a performance similar to the outputs. It appears that not all the participants remained involved due to the fact that the 'traveler from the future' was too far much ahead and introduced large amounts of information and claims in a short time. It would be interesting to investigate how the contributions of 'travelers from the future' can be optimally exploited: regarding team process performance as well as the performance on service definition outputs.

Fourthly, an area for future research is the applicability of MuCh-QFD to other auxiliary services. As stated in section 8.2, we expect that MuCh-QFD can be applied not just to other ICT-enabled services, but even to new 'personal contact' services that have to fit into an overall channel mix. It would be interesting to repeat our field experiment with other types of services and test the robustness of our findings. If it turns out that MuCh-QFD is robust in various different contexts this has great practical value for the numerous innovations that are constantly taking place and that could use some support in making customer-oriented choices that are sensitive to other channel services (which are potentially competing, but also potentially complementary).

The second topic of this section is design research methodology, which is still a young and underdeveloped, but also very relevant discipline (Hevner, March, Park, Ram, 2004; Aken, 2005). We think that our field experiment illustrates a promising experimentation paradigm for the development and testing of design methods. Our research shows that the approach explained in section 1.3, combining design sciences with empirical traditions from social sciences, is academically fruitful in two ways. Firstly, the performance of different design methods can be tested experimentally in relatively efficient and robust ways if those methods can be represented in standardized formats of half- or full day design sessions. Secondly, the pressure to translate methods into short session formats that enable practical experimental testing stimulates the development of new design methods and new insights into which design tasks are most crucial in which design phase. In our research, for example, our method evolved significantly between the first and second pre-test, and our final method is much more condensed than the first version. Also, insights into why a specific design task adds value, and in combination with which other design tasks, become more precise due to the experimental setting and due to observing the effects of the same design tasks for five or six cases. Our experiences confirm the statement that after testing with five users (or with five cases in our research) additional cases add few additional insights: in usability testing the first five users uncover 85% of usability problems; number six uncovers about 5% additional problems and after nine users/cases 95% of problems have been uncovered (Nielsen and Landauer, 1993; Nielsen, 2004b).

There are several challenging questions for this area of research. For example: How to develop control group design methods? How to test experimental design methods against control group methods? How to translate methods into session formats (and maybe other forms suitable for efficient testing with multiple cases)? And what is the external validity of these types of field experiments? How to develop measurement instruments that can measure the quality of design outputs? And how to create design process quality measures that are generally applicable? In our research we have formulated preliminary answers to several of these questions, but we feel that many more options exist that can be fruitful. Moreover, we hope that a healthy and growing body of knowledge on design research and design research methodology can help make design methods and their value more transparent to others than those who are experts in those methods.

Appendices

Appendix A: Explanation of QFD

Appendix B: Intake protocol

Appendix C: Session agendas in GroupSystems™

Appendix D: A reason to test with only five or six users

Appendix E: Measurement instruments

Appendix F: Case summaries

Appendix G: Observations and scores per session

Appendix H: Session output analyses on outcome quality requirements

Appendix I: Statistical details for section 7.2.4 - Disturbing factors

Appendix J: Abbreviations

Appendix A: Explanation of QFD

What is Quality Function Deployment (QFD)?

Quality Function Deployment (QFD) was developed in Japan in an effort to encourage engineers to consider quality early in the design process (Mizuno & Akao, 1994). 'Quality Function Deployment' is a direct translation of the characters *Hin Shitsu, Ki No, Ten Kai*, which means something like the strategic arrangement (deployment) throughout all aspects of a product (functions) of appropriate characteristics (qualities) according to customer demands. Some of the key words to capture the spirit of QFD are: attaining high quality, preventing design flaws, stimulating communication across the organization and throughout the product life cycle. As Clausing (1994) stresses, it is important not to treat QFD merely as a set of matrix-filling exercises. The underlying philosophy is at least as important. Also, when this method is only used in one department of an organization (for example in design), and not throughout other departments as well, it is bound to be less effective than it can be. It represents a way of innovating, rather than just a simple design tool.

QFD started in the Kobe shipyards as a way to expand and implement the ideal of quality as it was perceived at the time. From there it was developed much further by the Japanese automotive industry. Toyota in particular used it to reduce development time significantly. It was very successful in reducing the required change orders after production was started. QFD has been one of the keys to the company's success.[19] QFD has been evolved by product development people in response to the following major problems that were identified in the traditional process (Clausing, 1994):

- Disregard for the voice of the customer
- Disregard for the competition
- Concentration on each specification in isolation
- Low expectations
- Little input from design and production people into product planning
- Divergent interpretations of the specifications
- Lack of structure
- Lost information
- Weak commitment to earlier decisions

QFD is a systematic, matrix-based, visual approach to design quality products and services. It is based on the Total Quality Development philosophy as described by Clausing (1994), which states that high-quality products and services distinguish themselves by adhering to quality standards throughout their life cycle. The specification of quality requirements and the deployment of quality for such products and services begins as early as possible in the life cycle. Furthermore, the quality requirements are obtained directly from the customers.

In section 4.1 we explained the QFD 'House of Quality', which forms the set of matrices that provide the high level description of the design (linking customer priorities to functions, competitor performance, technical benchmarks and performance standards). In our research we do not move below this high-level

---

[19] http://www.proactdev.com/pages/ehoq.htm

service description. However, in chapter 8 we mention how QFD originated from a need to integrate design development and implementation phases into design decisions in the initiation phase. Hence, we briefly explain below how QFD links detailed engineering questions to the specifications and customer needs of the House of Quality.

The way to carry the prioritized information from the matrices used in QFD throughout the development process, is to deploy the matrices in a hierarchical fashion, in the form of several matrices linked with respect to vertical and horizontal output. This means that the columns in one matrix become the rows of the next matrix in order to be correlated with more detailed information in the columns, which will serve then as the next matrix' row input, and so on (Herzwurm et al., 2002). This is illustrated in Figure A-1.

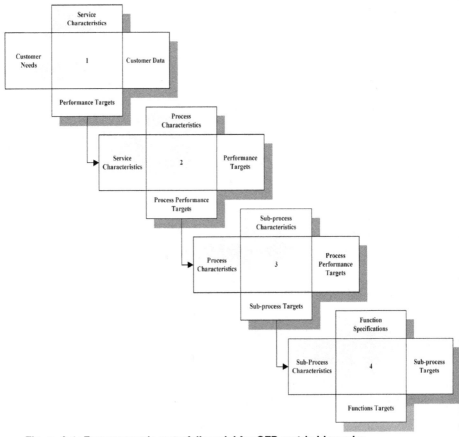

Figure A-1: Ramaswamy's waterfall model for QFD matrix hierarchy.

Different conventions exist in the QFD literature to represent the hierarchy of matrices. Ramaswamy (1996) describes a total development process customized for service design. The figure shows the most common convention, referred to as

174

the American Supplier Institute (ASI) model or Clausing Model, which Ramaswamy has customized for service design. The first matrix represents rooms 1 to 4 and 8 of the house of Quality (what Ramaswamy calls service characteristics is what we call service functions). In the second level matrix labeled 2, which Ramaswamy calls the 'service/process matrix', the service characteristics and performance targets of matrix level one are taken as inputs. On the basis of these, process characteristics and performance targets are defined. These are then used as input for level three, where sub-processes and – targets are defined. In level four, detailed specifications for hardware, software, documentation, training etc are generated to meet the sub-process performance requirements of level three. This illustrates how, by linking the output of each matrix to the input of the following matrix, the hierarchy allows the voice of the customer to determine the design of the service down to the most detailed level.

Rules of thumb for filling in a relationship matrix with correlation scores
Finally, we discuss the QFD rules of thumb for filling a relationship matrix with correlation scores. The extent to which a function contributes to each customer need can be inserted into the cells of a relationship matrix. A '9' indicated when a correlation is strong, a '3' when a correlation is moderate. If necessary, a '1' can be awarded to indicate a weak correlation, but this can often be left out, since it adds limited value to the discussions.

Table A-1: Rules of thumb for filling in the relationship matrix:	
1.	All the needs in the matrix are met by at least one function
2.	Identify at least one function with a strong correlation for every need
3.	One single function covering all the needs does not imply a flexible design. Identify approximately as many functions as there are needs, i.e. the 9's fall approximately along the diagonal of the matrix
4.	No more than one third to one half of the cells should be filled (a few important relationships should shape the design; relating every function to every need would draw attention away from these few important relationships)

In the table we indicated the rules of thumb as taken from Ramaswamy (1996). We explained them to the participants and monitored whether they adhered to these rules during our sessions. We also followed a practical guideline to fill in the matrix, which was to start by identifying all the nines. Some of them were then examined more closely, and changed into threes when necessary (Afterwards, additional threes were identified in the remainder of the discussion) . The advantage of this procedure is that it helps people focus on the main correlations, making it easier to decide what the actual value of the various correlations is.

Appendix B: Intake protocol

Intake protocol and questions

Introduction:
- First of all we want to thank you for your interest and for taking part.
- Secondly, why are we here: for us, this meeting has two purposes. First of all, it is a test to see whether there is a match between your service idea and our method. After all, it is possible that you have a wish that cannot be facilitated with our method, in which case we will not continue with a session. The second goal is to use this meeting as a starting point for a successful session, which is why we will collect as much input as possible.
- Thirdly, it is good to tell you a little more about the process. After this intake we sometimes hold a second intake meeting to cross some t's and dot some i's. After that we conduct the session, and finally the results of the session are presented to you and the team. Needless to say, your information will be handled in the utmost confidence.

Description Product-Market Combination & desired Internet service
1. What is the intended online service? (add detail in final question)
   a) Initial wording:

   b) Wording after completion of the remainder of the interview:

2. What is the product & what is the intended target group?
   a) Product:

   b) Target group:

Customer process & what must be the focus for success?
3. What is the process that the end-user goes through:
   a) What are the 3 or 4 main steps from pre-sales (orientation) to after-sales (use, incidents, mutations, etc)?

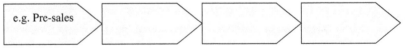

e.g. Pre-sales

   b) What are the 3 or 4 sub-steps per main step (for instance, comparing products in pre-sales)

1)	1)	1)	1)
2)	2)	2)	2)
3)	3)	3)	3)
4)	4)	4)	4)

4. How mature or innovative does the customer think the product is?
[Answer on the basis of the Market Life Cycle Model of Lynn (2000).
Auxiliary questions: a) Who are the two main competitors, and on the basis
of what does the customer choose between one of them and you? b) Is it a
'high' or 'low support' purchase?] -> In what phase are customers (I to IV)?

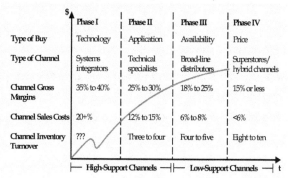

	Phase I	Phase II	Phase III	Phase IV
Type of Buy	Technology	Application	Availability	Price
Type of Channel	Systems integrators	Technical specialists	Broad-line distributors	Superstores/ hybrid channels
Channel Gross Margins	35% to 40%	25% to 30%	18% to 25%	15% or less
Channel Sales Costs	20+%	12% to 15%	6% to 8%	<6%
Channel Inventory Turnover	???	Three to four	Four to five	Eight to ten

├─ High-Support Channels ─┼─ Low-Support Channels ─┤ t

5. What sub-step (and between which main step) of the customer process is
the most important one for the commercial success of this PMC (product-
market combination)? Do we choose this sub-step as the focal point for the
service design?

Service online, offline and at competitors
6. What Internet services & functionality are currently available to customers for
this sub-step?
a)
b)
c)
d)

7. What Internet services and functionalities for customers are being offered by
the two main competitors?
Competitor 1:
a)
b)
c)
d)
Competitor 2:
a)
b)
c)
d)

8. What services & functionality are you offering offline for this sub-step?
a)
b)
c)
d)

9. Which of the offline functions mentioned above could be interesting to some customers over the Internet? And why?

10. Which of the above-mentioned offline services offer added value to customers compared to Internet services? And why?

Costs and benefits
11. What are the costs involved in setting up the Internet service?
    a) How much time & labor costs for consultation between parties involved?
    b) How much time & labor costs for design and implementation?
    c) How much time & labor costs for annual maintenance and management?
    d) What are the costs for software, hardware and possibly hosting?

12. What additional revenues or cost savings do you envisage if the Internet service is a success?
    a) per customer
    b) how many customers will be using this service in your prediction after one year and after two years?
    c) So the revenues in two years time are:

Participants & agreements
13. Which participants do you think should be part of the team? -> see form below

14. What date would you choose? -> see form below

15. Is there any aspect of the Internet service description you feel should be reformulated?   - > see question 1

16. Finally: a number of students will be filling in an observation protocol during the session. Is that alright with you?

Date:

Team & initiator:

Participants:

Customer focus:	Customer contact focus:	Marketing focus:	Operations & IT:
1.	1.	1.	1.
2.	2.	2.	2.
Reserve:	Reserve:	Reserve:	Reserve:

Appendix C: Session agendas in GroupSystems™

MuCh-QFD Agenda (morning session starts at 8.00, afternoon session at 13.00):

Agenda item 0
8:00   Introduction
8:15   Presentation of intake results
8:30   Exercise (Topic Commenter)

Agenda item 1
8:35   What are customer needs and what are possible web functions? (Topic Commenter)
8:45   What are the main needs in this buying phase? (Categorizer)
8:55   What are the main functions for this buying phase? (Categorizer)
9:05   What are the main customer needs? (Vote)
9:10   If necessary: what are the Top 10 functions? (Vote)

9:15   Break

Agenda item 2
9:25   Presenting service matrix: needs and functions
9:30   Breakout: team work
    a) Drawing the service matrix: how are the functions related to the needs? [10 min]
    b) What is the service slogan & e-service concept? [5 min]
    c) What are the solutions that can be offered per function? [10 min]
    d) Entering the results in an Excel sheet [5 min]

Agenda item 3
10:00   When does the customer prefer other channels (in this buying phase)? (Topic Commenter)
10:10   What tasks of the other channels are the most important? (Vote)
10:15   What are possible win-win situations between channels? (Topic Commenter)
10:25   What are the top 3 win-win situations? (Vote)

10:30   Break

10:40   Breakout: team work
    a) Expand the service concept by adding tasks from other channels in this buying phase (including needs) [15 min]
    b) Which new solutions to offer per function? [15 min]

Agenda item 4
    c) What are the strengths and weaknesses compared to competitors and old situation? [10 min]

11:20   How do the new customer needs score? (Vote)
11:22   How does the IST (current site) score on customer needs? (Vote)
11:25   How does competitor 1 score on customer needs? (Vote)
11:28   How does competitor 2 score on customer needs? (Vote)
11:30   How does the new e-service concept score on customer needs? (Vote)
11:35   Discussion

11:50   Questionnaires and evaluation

FE agenda (morning session starts at 8.00, afternoon session at 13.00):

Agenda item 0
8:00 AM   Introduction
8:15 AM   Presentation of intake results
8:30 AM   Exercise round (Topic Commenter)

Agenda item 1
8:35 AM   What are the goals of the e-service? (Topic Commenter)
8:45 AM   Clustering the goals into general categories (Categorizer)
9:00 AM   What are the most important goals? (Vote)

9:05 AM   Break

Agenda item 2
9:20 AM   What are the web functions for each main goal?? (Topic Commenter)
9:35 AM   Making set of main functions (Categorizer)
9:55 AM   What are the most important web functions? (Vote)

10:05 AM   Break

Agenda item 3
10:20 AM   Breakout: team work
                a) What is the service slogan & e-service concept? [10 min]
                b) Select a number of solutions that form the e-service concept [25 min]
                c) Is this service good enough to warrant investment, and why? [10 min]
                d) Present result a) b) c) [5 min]

Agenda item 4
11:10 AM   How does the e-service concept score in relation to the goals? (Vote)
11:20 AM   Discussion

11:40 AM   Questionnaires and evaluation

Appendix D: A reason to test with only five or six users

Here we illustrate why testing with a limited number of users (or cases in our research) helps to uncover most of the relevant flaws and strength of a design. This appendix is to a large degree based on the work of Jacob Nielsen and several others in the area of usability testing. The great thing about their work is that they have tested empirically and under various conditions how much new information becomes available each time an additional user is tested (Nielsen and Landauer 1993; Beyer and Holtzblatt 1998; Nielsen 2004a, 2004b; Norman 2004; Tullis and Wood 2004).

In 1993, Jacob Nielsen and Tom Landauer showed that the number of usability problems found in a usability test with n users is:

$$N(1-(1-L)^n)$$

where N is the total number of usability problems in the design and L is the proportion of usability problems discovered while testing a single user. The typical value of L is 31%, averaged across a large number of projects Nielsen and Landauer studied. Plotting the curve for $L=31\%$ yields the following result:

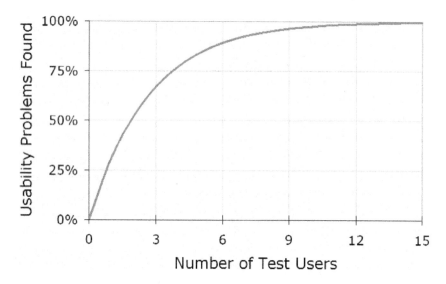

In the words of Nielsen (2004b) this graph indicates the following:
> *The most striking truth of the curve is that **zero users give zero insights**.*
> *As soon as you collect data from a **single test user**, your insights shoot up and you have already learned almost a third of all there is to know about the usability of the design. The difference between zero and even a little bit of data is astounding.*

*When you test the **second user**, you will discover that this person does some of the same things as the first user, so there is some overlap in what you learn. People are definitely different, so there will also be something new that the second user does that you did not observe with the first user. So the second user adds some amount of new insight, but not nearly as much as the first user did.*

*The **third user** will do many things that you already observed with the first user or with the second user and even some things that you have already seen twice. Plus, of course, the third user will generate a small amount of new data, even if not as much as the first and the second user did.*

*As you **add more and more users, you learn less and less** because you will keep seeing the same things again and again. There is no real need to keep observing the same thing multiple times, and you will be very motivated to go back to the drawing board and redesign the site to eliminate the usability problems.*

After the fifth user, you are wasting your time by observing the same findings repeatedly but not learning much new.

As a consequence, iterative design is preferred (Nielsen, 2004b):
You want to run multiple tests because the real goal of usability engineering is to improve the design and not just to document its weaknesses. The ultimate user experience is improved much more by three tests with 5 users than by a single test with 15 users. After the first study with 5 users has found 85% of the usability problems, you will want to fix these problems in a redesign.

After creating the new design, you need to test again. Even though I said that the redesign should "fix" the problems found in the first study, the truth is that you think that the new design overcomes the problems. But since nobody can design the perfect user interface, there is no guarantee that the new design does in fact fix the problems. A second test will discover whether the fixes worked or whether they didn't. Also, in introducing a new design, there is always the risk of introducing a new usability problem, even if the old one did get fixed.

Also, the second test with 5 users will discover most of the remaining 15% of the original usability problems that were not found in the first test. (There will still be 2% of the original problems left - they will have to wait until the third test to be identified.)

Although we set up our research design independently from these guidelines, the similarities are striking: we conducted 3 rounds of pre-tests before starting our final experiment, and indeed most adjustments to our method were made on the basis of the first pre-test, and the second pre-test in turn led to more adjustments than the third pre-test.

Appendix E: Measurement instruments

In this appendix we provide:
1. The background and opinion questions for MuCh-QFD session questionnaires
2. The background and opinion questions for FE questions that were different from MuCh-QFD questionnaire questions
3. The objective events questions (items per design requirement)
4. The observation protocols used for agenda parts I to IV
5. Our original constructs and items and an explanation of changes made

1. Background and opinion questions MuCh-QFD sessions
Each participant was asked several questions regarding background and opinions. These were used to check the impact of several potentially disturbing factors. Moreover, in the interview immediately following the survey, the subjective questions encouraged them to provide their opinions, and helped us interpret their remarks.

I – Background questions
1. Personal:
    a) What is your name?
    b) What is the name of your company?
    c) What is your position at the company?
    d) How long have you been working in this sector?
    e) In a next meeting the results of this workshop will be presented. Are you interested in attending that meeting?
2. In what role did you take part in the workshop?
         Customer focus / Intermediary / Insurer / Operations & IT / Other, namely:
3. Did you take part before in a workshop to define a new service?
         Yes / No
4. Have you been involved before in defining a new service that is somewhat comparable to the service that has been described today?
         Yes / No

5. Please evaluate the following statements:
    a) This session has lived up to my expectations
         Disagree strongly / Disagree / Neutral / Agree / Agree very much
    b) When I include my experience with other design methods, I am satisfied with the approach
         Disagree strongly / Disagree / Neutral / Agree / Agree very much
    c) The session has the required level
         Disagree strongly / Disagree / Neutral / Agree / Agree very much
    d) The structured preparation of the workshop via the intake was useful
         Disagree strongly / Disagree / Neutral / Agree / Agree very much
    e) This workshop is helpful in defining an e-service more quickly
         Disagree strongly / Disagree / Neutral / Agree / Agree very much

6. Did you expect prior to the workshop that it might help you define the service quickly?
         Certainly not / Probably not / Neutral / Probably / Certainly

7. Do you think you will use the results of this workshop in the future?
         Certainly not / Probably not / Neutral / Probably / Certainly

8. Please evaluate the following statements:
    a) I have become more aware of the various dilemmas involved in defining this service
        Disagree strongly / Disagree / Neutral / Agree / Agree very much
    b) Innovation in our sector is a cumbersome process.
        Disagree strongly / Disagree / Neutral / Agree / Agree very much
    c) The approach of defining a service together with other stakeholders helps initiate successful new services.
        Disagree strongly / Disagree / Neutral / Agree / Agree very much
    d) A careful service definition helps initiate successful new services.
        Disagree strongly / Disagree / Neutral / Agree / Agree very much

II – Evaluation of the workshop, incl. introduction
9. Which parts of this workshop did you feel were the *most* useful? Select 5 at the most, and briefly explain why.
Part I
a) Identifying the most important customer needs
b) Identifying the most important Internet functions
Part II
c) Using a matrix to evaluate functionalities in relation to customer needs
d) Defining service slogan
e) Identifying core solutions for the site
Part III
f) Checking whether customers may appreciate support from other channels
g) Looking for win-win relationships between channels
h) Extending the matrix with functions and solutions from other channels
Part IV
i) Comparing the new e-service with the existing solution and with competitors' sites

10. Which parts of this workshop did you feel were the *least* useful? Select 5 at the most, and briefly explain why.
Part I
a) Identifying the most important customer needs
b) Identifying the most important Internet functions
Partl II
c) Using a matrix to evaluate functionalities in relation to customer needs
d) Defining service slogan
e) Identifying core solutions for the site
Part III
f) Checking whether customers may appreciate support from other channels
g) Looking for win-win relationships between channels
h) Extending the matrix with functions and solutions from other channels
Part IV
i) Comparing the new e-service with the existing solution and with competitors' sites

11. What do you think of the workshop as a whole?
        Very useful / useful / neutral / relatively useless / very useless
        How would you rate the workshop on a scale from 1 to 10?

12. How useful do you find the workshop to:
a) make the design customer-oriented?
        Very useful / useful / neutral / relatively useless / very useless
b) arrive at a good integration between the website and the support via other channels (telephone, face-to-face, etc)
        Very useful / useful / neutral / relatively useless / very useless

c) help save costs or enhance revenues at other channels?
        Very useful / useful / neutral / relatively useless / very useless
d) improve competitive position via e-services?
        Very useful / useful / neutral / relatively useless / very useless
e) make relatively good progress in designing an e-service in a short time?
        Very useful / useful / neutral / relatively useless / very useless
f) give focus to an e-service design?
        Very useful / useful / neutral / relatively useless / very useless
g) improve interaction and communication between various perspectives (customer, intermediary, insurer, operations & IT)?
        Very useful / useful / neutral / relatively useless / very useless

13. How efficient was the workshop in your opinion?
        Not efficient at all / not very efficient / average / efficient / very efficient

14. Did all the participants contribute equally to the process?
        Not equally at all / not equally / neutral / equally / very equally

15. Did you feel you were able to express your ideas, wishes and concerns?
        Very badly / Badly / Neutral / Well / Very well

2. Background and opinion FE questions different from MuCh-QFD

Since the agenda items of FE and MuCh-QFD sessions were different, questions 9 and 10 had to be changed for the FE session survey. The adapted questions are displayed below.

9. Which parts of this workshop did you feel were the *most* useful? Select 5 at the most, and briefly explain why.
Part I
a) Identifying goals
b) Prioritizing goals
Part II
c) Identifying e-service functions
d) Prioritizing e-service functions
Part III
e) Defining service slogan
f) Identifying core solutions for the site
g) Discussing whether this e-service is good enough to warrant investment
Part IV
h) Re-evaluating the priorities of goals
i) Rating the e-service concept in relation to the goals

10. Which parts of this workshop did you feel were the *least* useful? Select 5 at the most, and briefly explain why.
Which parts of this workshop did you feel were the *most* useful? Select 5 at the most, and briefly explain why.
Part I
a) Identifying goals
b) Prioritizing goals
Part II
c) Identifying e-service functions
d) Prioritizing e-service functions
Part III

e) Defining service slogan
f) Identifying core solutions for the site
g) Discussing whether this e-service is good enough to warrant investment
Part IV
h) Re-evaluating the priorities of goals
i) Rating the e-service concept in relation to the goals

3. Questions regarding the occurrence of certain events

The 'objective events' questions measure the performance of our sessions in terms of our design support requirements. These requirements stem from literature and observations from our multi-channel cases (see also chapter 3). The first seven requirements are applicable to the definition phase we address in our workshop. The eighth requirement, communication of concept coherence during implementation, lies outside the scope of our e-service definition sessions.

1. Customer-oriented design	
a.	Did the teams discuss what the 5 main customer needs are?	y / n
b.	Have the customer needs been prioritized?	y / n
c.	Have the customer needs been verified explicitly with customers?	y / n
d.	Have customer needs been used explicitly as a starting point for the design?	y / n
e.	Have the choices between the various e-service options been linked to the customer needs?	y / n
f.	Has been checked whether the e-service designs add something to the existing services on the market? (In terms of higher scores on customer needs than competitors)	y / n

2. Value and coherence of channels	
a.	Did the team discuss the existing services: - that are already offered via the Internet? - that are offered via other channels?	y / n y / n
b.	Did the team discuss the customer-related pros and cons of the other channels?	y / n
c.	Did the team design an overall solution that explicitly combines the advantages of the other channels?	y / n
d.	Did the team discuss 'handovers' between channels (to ensure that there is a service flow between the channels)?	y / n
e.	Did the team discuss whether providing information on the Internet reduces the time needed to explain to the customer in person what a product is and what the benefits are?	y / n
f.	Did the team explicitly evaluate various possible service solutions?	y / n

3. Synergy between channels	
Synergy is achieved when channels reuse existing resources. Has it been discussed whether the new e-service:	
a.	- reuses existing customer relations?	y / n
b.	- reuses the existing reputation?	y / n
c.	- reuses existing logistical processes?	y / n
d.	- reuses existing information systems?	y / n
e.	- reuses existing information between channels?	y / n
f.	- automates supplier tasks (such as entering orders, sending documents, etc) via self-service?	y / n
g.	- reduces the need to train staff?	y / n

4. Match with marketing strategy & improved competitive position		
a.	Has the e-service been placed explicitly within the overall marketing strategy?	y / n
b.	Has it been explicitly discussed what competitors do with their e-service?	
	- 1 competitor	y / n
	- 2 competitors?	y / n
c.	Has it been explicitly discussed what competitors do in terms of their marketing strategy?	
	- 1 competitor	y / n
	- 2 competitors?	y / n
d.	Has an explicit evaluation been made where to copy what competitors do and where not?	y / n
e.	Has it been verified whether the e-service matches the overall marketing concept (for instance: brand, customer segments, services of other channels, the core product, marketing communication, etc)?	y / n
f.	Has it been verified hoe likely it is that the new service will generate or win back customers (in comparison to competitors)?	y / n
g.	Has it been verified whether the e-service generates financial benefits?	y / n

5. Progress during the session		
Which aspects of the e-service design have been discussed?		
a.	- customer needs?	y / n
b.	- value of the e-service for the customer compared to competitors?	y / n
c.	- the relationship between the business strategy and the way the e-service has to offer added value to customers?	y / n
d.	- other customer contact points in addition to the website?	y / n
e.	- the way channels work together in offering the service?	y / n
f.	- generating alternative service solutions?	y / n
g.	- refining service solutions?	y / n
h.	- evaluation of technical (im)possibilities?	y / n

6. Focus design process		
a.	Did the team make decisions as a whole?	y / n
b.	Was it clear to everyone what items had to be discussed?	y / n
c.	Was it clear to everyone what the eventual results had to be?	y / n
d.	Were the discussions structured?	y / n

7. Communication between stakeholder interests in design team		
a.	Did the team agree on customer needs as an important starting point?	y / n
b.	Did the team agree on the consequences of customer needs for the design of the e-service?	y / n
c.	Did the team agree on the competitive position as an important starting point?	y / n
d.	Did the team agree on the consequences of competitive position for the design of the e-service?	y / n
e.	Did the team agree on the service provisioning via other channels as an important point of reference?	y / n
f.	Did the team agree on the consequences of the services provided via other channels for the design of the e-service?	y / n

4. Observation protocol (both sessions, agenda parts I to IV)

As explained in chapter 6, the observation protocol was developed and tested as a condensed version of our 'objectified' questions.

Observation Protocol - Date:

Part: Agenda-item 0, 13.00 - 13.30

Name observer:

1. Customer oriented design and value creation

	Five main customer needs discussed
	Customer needs prioritized
	Customer needs linked to design choices
	Scoring design on customer needs compared scoring competitors
	Number of checks

2. Channel value and coherence

	Services offered via other channels discussed
	Services that are already being offered via the Internet discussed
	Pros and cons of other channels for customer discussed
	Discussed whether information n the Internet reduces time (for explanation, etc) in other channels
	Handovers between channels discussed
	Number of checks

3. Channel synergy

	Discussed whether the new e-service:
	- reuses existing customer relations?
	- reuses existing reputation?
	- reuses existing logistical processes?
	- reuses existing information systems?
	- reuses existing information between channels?
	- automates supplier tasks via self-service?
	- reduces the need to train staff?
	Number of checks

4. Competitive positioning

	Place e-service within marketing strategy discussed
	Competitors' e-services discussed
	Marketing strategy of competitors discussed
	Evaluated where to emulate competitors and where not
	Verified whether e-service matches supplier's role and image
	Verified whether the e-service will generate/win back customers
	Cost benefits e-service discussed
	Number of checks

5. Progress during session

	Has the desired objective of this part been realized?
	Has this part been result-oriented?
	Has this part been dealt with efficiently?
	Number of checks

6. Focused design process

	Does the team make decisions together?
	Is everybody aware of the relevant problems?
	Is it clear to everybody what end-products are demanded?
	Are the discussions structured?
	Number of checks

7. Communication between stakeholder perspectives

	Did the perspective of each participant become clear?
	Team agreed on customer needs?
	Team agreed on competitive position?
	Team agreed on importance other channels in service mix?
	Team agreed on consequences customer needs for the e-service?
	Team agreed on consequences competition for the e-service?
	Team agreed on consequences channel mix for the e-service?
	Number of checks

5. Description of original constructs

We started developing our measurement instruments by defining constructs per design requirement. How well a team fulfills a requirement depends on several issues. We made those issues explicit as items. Originally, we developed ten items per design requirement. The items have been formulated as yes/no questions and each 'yes' answer earns one point. The items were used as index scales for constructs, by adding up all 'yes' scores per construct. Hence in total ten points could be generated per construct. As described in paragraph 6.3, several items per construct had to be removed or changed in subsequent review and test rounds, before we arrived at our final measurement instruments. This section presents our original 10-item constructs. For the eighth requirement we also devised a set of measurement items, for reasons of completeness, even though, our measurements during sessions only address the service definition phase. The eighth construct is not used in our experiments.

This approach introduces the assumption that all items represent equal intervals in our constructs. Although we have attempted to formulate our items in such a way that this may be mostly true, we did not test this assumption. We used nonparametric tests next to interval-based tests, and observed that they generate the same significance levels. These nonparametric tests operate under the milder assumption that rank order testing can be performed, meaning that, for instance, a score of 4 is higher than a score of 3, even if those scores are generated via different items.

ad 1. Customer-oriented Design

This requirement means that the team is focused on generating value for customers. Many services provided by channels are 'auxiliary' (i.e. not in the core offer) and non-billable. This hampers straightforward (financial) value comparisons between the services. Nevertheless, they are valuable to customers. How can we incorporate value for customers in the channel service design? How do we decide which service features to include in the total service offer, and which not?

To create value for customers in an economic and lasting way means providing benefits that are expected to be sustainable for a certain period (i.e. costs that are made for auxiliary service provisioning have to be justifiable in terms of competitive position, and in terms of generating or retaining customers)[20]. This implies that the team must know what customer preferences are, second it must know what the offers of the primary competitors are, and how the new e-service design relates to these offers. Hence, a team must know whether an e-service adds something to what is offered in the market. (Finally, the team should check whether the costs associated with different e-service solutions are justified, but this item belongs to construct number 4, aimed at the competitive position of the supplier.)

---

[20] As a result, our definition of value creation is relatively narrow: we exclude the provisioning of costly services for free when there is no clear idea as to how to earn back those costs. Even though a certain value is created for customers in the short run, this is not sustainable. This type of value creation falls outside the 'normative structure' value propositions of channel systems (Bucklin, 1966; Bucklin, 1972). In other words: if a firm creates a cost structure for itself that can not be translated into a higher position on the price quality line in the market, it will erode the other quality it can provide for a given price, see also chapter 2. According to our definitions, this results in an overall value destruction instead of value creation.

Following QFD and total design procedures (Clausing, 1994; Cohen, 1995; Ramaswamy, 1996), all team members must know what the customer requirements are (a), and the priorities for customers (b). Generally, certain people in the organization will have a good idea of customer preferences, but they also have to be explicitly checked with the targeted customer segment (c). In terms of design process, the team must take these customer preferences as their starting point (d) and design a customer process aimed at a high level of service quality and customer friendliness (e). A team must link its e-service design choices consistently back to customer priorities (f) and the customer process design (g). Next, it has to be checked whether e-service designs add something to what is offered in the market, in terms of fulfilling customer preferences and the customer process (h). Finally, customer centricity should not only guide design details, but also the overall customer process of the resulting total service proposition (i).

a. Do all team members know and discuss what the customer requirements are? (1 point)
b. Are customer requirements prioritized? (1 point)
c. Are customer preferences lists explicitly checked with customers (1 point)?
d. Are customer preferences taken as an explicit starting point for design (1 point)?
e. Is an explicit customer process design made that is aimed at customer-friendliness[21] (1 point)?
f. Are choices between alternative e-service design solutions linked back consistently to customer priorities (1 point)?
g. Are choices between alternative e-service design solutions linked back consistently to user-friendliness of the customer process design (1 point)?
h. Is checked whether e-service designs add something to what is offered in the market?
   - in terms of higher scores on customer preferences than competitors (1 point)?
   - in terms of a user-friendly customer process than competitors (1 point)?
i. Are the customer process and user friendliness of the total service proposition (new e-service design and 'new' channel mix) explicitly checked for performance on customer priorities (1 point)?

(total = 10 points)

2. Channel Coherence

Different channels offer different (dis)advantages to customers. These (dis)advantages have to be taken into account when designing a channel mix where channels strengthen and complement each other. Moreover, customers are generally used to interacting with an organization in certain ways. The new e-service and channel mix should be a logical extension to or modification of the existing service mix for customers (a). Next, the team should consider what each channel has to offer (b), and the new service proposition should be forged such that it combines the advantages of different channels (c). Specifically for the Internet, there are often possibilities to provide information services to customers that can enhance the value that they get from the other channels (d). During service operations, it is important that service handovers between channels are well-managed and well-designed. The design team should explicitly consider which service handovers exist and how they should be handled (e). A next step is to design for a seamless experience[22] (f). Different (functional and technical) solutions are possible for

---

[21] We define customer-friendliness in a broad sense: ease of use and the Servqual dimensions (tangibles, empathy, responsiveness, reliability and assurance) are important.
[22] Definition of a seamless experience: 1) No delays between channels: for example data entered online is immediately known in the other channels when a customer contacts

the new e-service. Those solutions have to be compared to each other, preferably in quantitative terms (g). Finally, the overall channel mix service proposition has to be has to be easy to understand for customers and send one clear message of what the proposition is (h).

a. Does the team discuss[23] the existing service mix, as already provided through the Internet (1 point) and through other channels (2 points) in relation to the new e-service?
b. Does the team discuss the (dis)advantages of each channel in relation to the desired service proposition (1 point)?
c. Does the team design a total proposition that explicitly combines the advantages of the different channels (1 point)?
d. Does the team discuss the information provisioning potential of Internet services to enhance the value provided via other channels (1 point)?
e. Does the team discuss service handovers between channels (1 point)?
f. Does the team explicitly design for a seamless experience between channels (1 point)?
g. Does the team explicitly evaluate the different e-service solutions that are possible (qualitatively: 1 point; quantitatively: 2 points)?
h. Is the overall channel mix proposition, including the new e-service design, explicitly evaluated for sending one clear service proposition message to customers (1 point)?

(total = 10 points)

3. Channel Synergy

Synergies occur when channels can re-use existing assets. For our purposes, when a new e-service is introduced in an existing channel mix, we focus on the question how the e-service can re-use existing channel assets or improve efficiencies by clever channel combinations. There are several assets that can be re-used: First, existing customer relations (a), which avoids the expenses of building a new customer base. Secondly, the existing brand (b), which avoids the expenses of building a new brand. Thirdly, existing procurement processes (c), which means that procurement management can be more efficiently used and sometimes higher discounts can be negotiated. Fourthly, existing processes for physical distribution (d). Especially for e-commerce the costs of shipping products home can be relatively high, and it causes delays that customers not always want. When customers can pick up products at for example a retail store near by, this can save costs and sometimes even provide faster delivery. Fifthly, existing information systems can be re-used (e), which can be used to enrich existing data and which avoids duplication of data, of systems and of systems management. Sixthly, back office processes can become more efficient (f) when large amounts of additional transactions are generated (i.e. an increase of 20% or more). The question for the new e-service is whether it is expected to generate that many more transactions in the back office. Seventhly, re-use of information across channels (g), which can reduce the amount of

---

them. Or a retail customer card is immediately valid online as well. 2) 'Warm' handovers: a channel (often in the form of a service employee) actively hands the customer over to the next channel, which ensures service continuity. 3) The customer has to explain his wishes and/or provide his name & address data only once.

[23] Instruction for scoring [also applies to similar scoring items]: A team is not given a positive score if it does not get further than an incidental remark from a team member on a certain issue. The team really has to address the issue as a team and reach some kind of 'verdict'.

additional market research that needs to be done to improve customer understanding. An eighth synergy is the reduction of manual labor (h) for example via customer self-service, which can generate cost savings. Ninth is the possibility to reduce the need for employee training by re-using information provided by the e-services in other channels (i). Think, for example, of product- and price information and information on temporary special offers. The details that are stored online can also be checked by employees in call centers, retail stores or in the field sales force. Finally, e-services can also be designed to reduce the amount of customer education that other channels have to provide (j). A lot of product information and after sales support can be provided online, which means that customers are often much better educated when they finally contact other channels.

    a. Is the new e-service design positioned to leverage existing customer relations (1 point)?
    b. Is the new e-service design positioned to leverage the existing brand (1 point)?
    c. Is the new e-service design positioned to leverage existing procurement processes (1 point)?
    d. Is the new e-service design positioned to leverage existing physical distribution processes (1 point)?
    e. Is the new e-service design positioned to leverage existing information systems (1 point)?
    f. Is the new e-service design positioned to generate significant economies of scale (20% extra transaction volume or more) in the back office (1 point)?
    g. Is the new e-service designed to re-use information across channels (1 point)?
    h. Is the new e-service designed to reduce or automate manual labor via self-service (1 point)?
    i. Is the new e-service designed to reduce the need for employee training (1 point)?
    j. Is the new e-service designed to reduce the need for customer training (1 point)?

(total = 10 points)

4. Competitive Positioning

The starting point is a standing organization with a given channel structure, a given type of customers and a given set of organizational capabilities. Hence, a multi-channel solution that would suit one firm in a given market does not necessarily suit its competitors. How to maintain a balance between IST and SOLL? How to enable choices between which givens to change and which to keep, or even use them as a competitive asset or differentiator? To ensure competitive strength, new e-service designs have to be positioned explicitly in an overall marketing strategy (a). Furthermore, the team has to know what the top 3 competitors are doing with their e-services (b) and their marketing strategy (c). Next, an explicit evaluation has to be made of what to emulate and what to do differently (d). Consequently, new e-service concepts should be evaluated on a number of aspects: Does the e-service fit into the overall marketing proposition (e.g. image and brand, customer segments, other channel services, core offer, marketing communication, etc) (e)? Does the e-service generate market value by generating or retaining customers? This can be assessed by evaluating e-services design scores on customer preferences (f). Does the e-service generate (long term) cost advantages (g)? And does the e-service provide value for customers at a value/cost ratio that is better than that of competitors (h)?

    a. Is the e-service explicitly positioned in an overall marketing strategy (1 point)?
    b. Is it explicitly mentioned what the top 3 competitors are doing with their e-services?
        – (Only one or two competitors: 1 point; at least the top 3 competitors: 2 points)

c. Is it explicitly discussed what competitors are doing in terms of marketing strategy?
   – (Only one or two competitors: 1 point; at least the top 3 competitors: 2 points)
d. Is an explicit evaluation made of what to do similarly and what differently (1 point)?
e. Is it checked whether the e-service fits into the overall marketing proposition (e.g. image and brand, customer segments, other channel services, core offer, marketing communication, etc) (1 point)?
f. Is it checked whether the e-service generates market value by winning or retaining customers, via assessing scores on customer preferences in relation to competitors (1 point)?
g. Is it checked whether the e-service generates cost advantages (1 point)?
h. Is checked whether the e-service provides value for customers at a value/cost ratio that is better than that of competitors (1 point)?

(total = 10 points)

5. Speed (-> Interpreted as 'Progress': How far does a team get in a session?)

In real life situations, service concept exploration only takes a limited period of time, typically a few weeks. In our experimental set-up, we have chosen a session format to compare design support methods against each other. In itself, a session is already a relatively fast method. Moreover, our experimental and control groups both follow sessions that take equally long. Hence, we interpret our measurement of progress on the speed requirement as: How far do teams get in the design session?

In terms of progress, an important element for group discussions is that the group reaches 'closure' on certain issues. This means that issues are first explored (divergent movement of discussion) and then some kind of conclusion is reached (convergent movement). Our focus is on the conceptualization phase of a new e-service. In this phase, progress means that all relevant aspects and stakeholders are considered (customers, supplier, other channels, technology, competitors etc), and a first evaluation of e-service concepts is carried out. Considering these items in later phases is generally more costly, and it is also more difficult to readjust the design and the focus of the people involved. A team needs to reach conclusions on several items: customer preferences (a), competitive position and -strategy (b), consequences of the competitive strategy for the main service features and customer preferences that are targeted with the new e-service (c), the service blueprint, i.e. what is the customer process, and what are the delivery processes that need to be in place (d), discussion of how service elements are delivered across channels (e), generation of alternative service solutions and concepts (f), evaluation of alternative service solutions and concepts (g) and evaluation of technical (im)possibilities (h).

Finally, another element is that the outcomes of the session have to be sufficiently meaningful and unambiguous to enable further detailing or decision making (i).

a. How many aspects of the e-service design were discussed in a way that provided closure for the team
   - customer preferences (1 point)?
b. - competitive position and –strategy (1 point)?
c. - consequences of the competitive strategy for the primary features and customer preferences that are targeted with the new e-service design (1 point)?
d. - service blueprinting (1 point)?
e. - discussion of how service elements are delivered across channels (1 point)?

f. - generation of alternative service solutions and concepts (1 point)?
g. - evaluation of service solutions and concepts (1 point)?
h. - evaluation of technical (im)possibilities (1 point)?
i. Are the outcomes of the session useful (meaningful and unambiguous) for further detailing or decision making (at least 50% of the outcomes: 1point; at least 80% of outcomes: 2 points)

(total = 10 points)

6. Focussed design process

The number of design space options explodes when discussing multi-channel service combinations and possible total service offers. How to manage/structure the design process? Moreover, different stakeholders, perspectives and priorities are present in the team. Hence the natural focus of attention of different team members will move into different directions. What is needed is a shared goal and a shared process. This can be described in more detail as: all team members participate in the same process steps (a), they discuss all major decision issues together (b), it is clear to all which issues should be dealt with (c) it is clear to all which intermediate results (d) and final results (e) should be delivered, it is clear to all that close cooperation is needed, and why this is important for the quality of the e-service design (f), discussions are efficient and aimed at (intermediate) results (g), discussions which generate insufficient progress on (intermediate) results for 3 minutes are stopped and put aside for future discussion after the session (h), and explorations (divergent idea generation) are followed by converging activities - leading to results in the form of short lists, priorities, decisions etc

a. Do all team members participate in the same process steps (1 point)?
b. Does the team discuss all major decision issues together (1 point)?
c. Is it clear to all which issues should be dealt with (1 point)?
d. Is it clear to all which intermediate results should be delivered (1 point)?
e. Is it clear to all which final results should be delivered (1 point)?
f. Is it clear to all that close cooperation is needed, and why this is important for the quality of the e-service design (1 point)?
g. Efficient discussions: Are team discussions clearly aimed at (intermediate) results (1 point)?
h. Efficient discussions: Are discussions which generate insufficient progress on (intermediate) results for 2 or 3 minutes stopped and put on an agenda to be addressed outside of the session?
    - no debates longer than 3 minutes that go very deep on one specific item or detail (1 point)?
    - no debates longer than 3 minutes that remain inconclusive / undecided (1 point)?
i. Are explorations (divergent idea generation) followed by converging activities - leading to short lists, priorities, decisions etc (1 point)?

(total = 10 points)

7. Communication between stakeholder perspectives in development team

A wide range of stakeholders is involved, ranging from CxO's to specialists on channels, IT, and sales and service processes. How to deal with the wide range of interests and support different perspectives? The different team members need to understand and appreciate all relevant aspects of the design (at least at an abstract level). All team members must understand why customer priorities are an important starting point for design (a) and what the implications are of the customer priorities on the e-service design (b). The same is true with regard to the importance and consequences of marketing strategy (c) and (d), service blueprinting (e) and (f), technical (im)possibilities (g) and (h), and other channel services (i) and (j).

a. Do all team members understand why customer priorities are an important starting point (1 point)?
b. Do all team members understand the implications of the customer priorities on the e-service design (1 point)?
c. Do all team members understand why marketing strategy is an important starting point (1 point)?
d. Do all team members understand the implications of marketing strategy on the e-service design (1 point)?
e. Do all team members understand why service blueprinting is an important starting point (1 point)?
f. Do all team members understand the implications of the service blueprint on the e-service design (1 point)?
g. Do all team members understand why technical (im)possibilities are an important starting point (1 point)?
h. Do all team members understand the implications of the technical (im)possibilities on the e-service design (1 point)?
i. Do all team members understand why other channel services are an important starting point (1 point)?
j. Do all team members understand the implications of the other channel services on the e-service design (1 point)?

(total = 10 points)

(8. Communication of concept coherence during implementation)

(Requirement 8 is excluded from our session experiment, but we also include requirement 8 to complete our measurement instrument (which firms may potentially use for self-assessment during implementations)).

During implementation, the balance between customer and supplier needs, and between one channel and the others, is easily lost. Especially when the design is handed over to other people and other departments. How can the coherence of the e-service design easily be maintained and communicated during implementation and operation? First, the design team must remain involved during the implementation, to enable communication and explanation where needed (a). Secondly, some manager must bare explicit responsibility for concept coherence during implementation and operation, with sufficient authority to lead the departments involved (b). Thirdly, the receiving organizations and departments must be provided with a clear, concise and coherent picture of what the e-service design is (c) and of why it is designed this way (d). Hence, possible changes during implementation can consistently be related back to: customer priorities, competitive position, the service blueprint, technical (im)possibilities, and the other channel services (e).

a. Does the design team remain involved during the entire implementation, including all stakeholder perspectives (1 point)?
b. Is there one manager y who explicitly has final responsibility for concept coherence, with sufficient authority to lead the departments involved (1 point)?
c. Are the receiving organizations and departments provided with a clear, concise and coherent picture of what the e-service design is (1 point)?

d. Are the receiving organizations and departments provided with a clear, concise and coherent picture of why the e-service is designed this way in relation to customer priorities, competitive position, service blueprint, technical (im)possibilities and other channel services?
(reference to 3 aspects or less: 1 point; reference to 4 or 5 aspects: 2 points)

e. Are implementation decisions and possible design changes due to the ongoing learning process consistently related back to:
- customer priorities, (1 point)
- competitive position, (1 point)
- service blueprint, (1 point)
- technical (im)possibilities, (1 point)
- other channel services (1 point)?

(total = 10 points)

Appendix F: Case Summaries

Case A	Absence management online for SME
Short Description	Online monitoring and management of employee absence (caused by illness, accidents, etc). Absence management is a complicated process due to many legal and financial obligations for employers. Transparency and manageability can be improved via Internet.
Customer Segment & Product	Small and Medium-sized Enterprises (SME's) with 5 to 50 employees, with a low risk of illness. The product is: insurance, plus absence management services. The e-service is: online process management.
Why?	Existing services are expensive, slow, involve too much risk and hassle, and offer too little control and choice too employers. (On the product side, absence management services can be provided faster and better tailored to SME needs, but this is outside our scope.) On the e-service side, processes can be made more transparent, reliable and tailored to the specific SME's needs.
Actors	Insurer, Insurance Agent, SME, Absence Mgt Service Provider
Customer Process Focus	Settlement and after-care phase, because insurance agents are given new tasks here (and so do the customers to some extent).
Main Internet Functions	Links to institutions; Process status & checklist; Notifications
Offline Functions	Accept and initiate damage process (by insurance agent); Advise in case of difficulties and help in online administration; Advice and treatment (= the absence management core product)
Business Case	For the insurance agent this is a new market, generating commission. For insurers this is a (relatively new) growth market. For the Absence Mgt Service Provider this is a profitable core product.
Case B	Insurer portal: Damage insurances
Short Description	This service should enable customers to buy and modify low-risk insurance policies via the intermediary website. The content of the site is provided/maintained by the insurer. It has the look and feel of the insurance agent.
Customer Segment & Product	The target group consists of the typical insurer consumers, who do not choose their policy primarily on price, but in the first place on comfort of doing business and the reliability of the insurer. The product is: a package of damage insurance policies that can be bought at a discount.
Why?	In the past there have been numerous initiatives by intermediaries to build websites, but most attempts failed due to lack of funds. This case offers a professional and low cost site.
Actors	Insurer, Insurance Agent, Consumers
Customer Process Focus	The presales phase. Within this phase a customer is provided with information about products, makes calculations and views quotations
Main Internet Functions	Advice tools (based on events or situation/profile); Making fee calculation; Giving information about the process
Offline Functions	Possibilities to buy the policy immediately when no PC is at hand; Negotiate; Handle more complex questions (e.g. regarding 'life' events)
Business Case	For the insurance agent and insurers it saves administration time and protects market share.

Case C	Intermediary portal: Damage insurances
Short Description	This service should offer a damage insurance portal that is supported by the intermediary sector and multiple insurers. It offers damage insurance policies via a central portal with the look & feel of the intermediary.
Customer Segment & Product	The target group consists of consumers that choose their insurance mainly on price. The product is: damage insurance policies.
Why?	Margins on the consumer market are currently so low that this is the only way to handle the competition from direct writers and Internet companies offering policies online.
Actors	Insurers, Insurance Agents, Developer(s) of the portal, Consumers
Customer Process Focus	Marketing & Use of the site: How to get the customer to visit the website, how to make sure he finds it easy to use?
Main Internet Functions	Product information; Calculate and compare offers; Advice on additional products; Giving information about the process; 'Contact me' functions
Offline Functions	Personal contact in case of 'life events' (multiple financial decisions), or due to habit (countered by price or other incentives to visit site). Assistance over the telephone for first time users
Business Case	The insurance agents' benefits are: saving time and possibly generating revenues. The insurers' benefit is saving on settlement costs
Case D	Digital registration of advice- and offering process
Short Description	This service should enable digital registration of the advice and quotation process of high-risk policies.
Customer Segment & Product	The target group consists of consumers. The product is: life insurance policies.
Why?	New legislation means stricter demands on independent insurance agents offering 'objective' advice. They have to demonstrate that the policy matches the customer's needs and resources. Also, he should be able to prove that the products of a sufficient number of insurers have been compared.
Actors	Insurance companies, Insurance Agents, System developers, Consumers, Government
Customer Process Focus	Pre-sales: orientation, advice and offering a policy.
Main Internet Functions	Information about products or offerings; Insight in the process; Insight in existing policies & conditions;
Offline Functions	Personal contact and advice in case of changes in personal situation, or changes in legislation, social security and/or taxation.
Business Case	Possibly quicker sales and more persuasive power, possibly less costs from government checks, quick reuse of previous advice

Case E	Online bill presentment
Short Description	Present the bill of mobile telephone users online. Offer the possibility to analyze this bill with some tools.
Customer Segment & Product	SME's up to 50 cell phone users. The product is: mobile communication (voice and data).
Why?	In the future customers will expect online bill presentment and analysis. This service helps customers control costs, and it is expected to reduce the rate of churn.
Actors	Mobile operator (marketing, customer contact, IT), Customers
Customer Process Focus	Usage phase, specifically billing & accounting.
Main Internet Functions	Personal portal; Analysis and reporting tools; Advice based on usage and bill; Data exports compatible with client systems.
Offline Functions	Personal contact for additional information and advice, negotiations, disputes, or reluctance regarding 'do-it-yourself'.
Business Case	Reduce churn rates.
Case F	Online telecom bundle offer, triple play
Short Description	This service offers triple play products and services online as a package.
Customer Segment & Product	The target group consists of consumers that choose products mainly on price and ease of doing business. The product is: Triple play: Television, Broadband Internet and Telephone.
Why?	It is possible to buy the TV product online, but customers that want TV should be triggered to buy Broadband Internet and Telephone as well in one package.
Actors	The triple play service provider (marketing, sales, IT), Customers.
Customer Process Focus	Orientation phase and sales phase.
Main Internet Functions	Product and price information; Advice based on personal situation and preferences; 'Contact me' functions
Offline Functions	Experience touch, look & feel in a shop. Personal contact for additional information and advice, or to aid buying decisions.
Business Case	Cross-sell from the Television offer to Telephone and BB Internet.

Appendix G: Observations and scores per session

Case A: Absence management

The absence management case was special in the sense that the 'core product' around which the e-services would be positioned was not yet ready and fairly complex, in contrast with all other cases, which involved existing and mainstream core products. This had a strong impact on the intake phase, which took about three times longer than average. The main challenge was creating focus and determining what to include in the session and what not to include. In terms of sessions output, the result of this was that the discussion that took place in Part III of the MuCh-QFD session focused on how tasks should be divided between parties involved, in addition to which suggestions were being made to change the core product - employee absence management. This illustrates that part III of MuCh-QFD helps people form a clear idea of what the customer-oriented tasks are. To some extent the focus on multi-channel issues distorted participant questionnaire scores with regard to channel synergy. According to our observations they mostly discussed coherence issues and to a lesser extent channel synergy: although this pattern is visible in the observer scores (upper two figures on the left), it is absent from the participant scores (bottom left). In the FE session defining the service slogan was relatively useful in comparison to other cases: due to the complexity of the service it helped to explicitly formulate its core purpose.

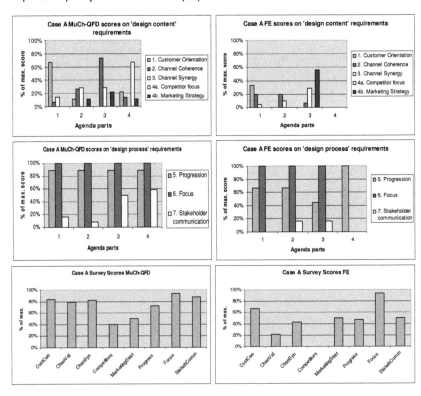

Case B: Intermediary portal

This was a corporate case, with a culture that was strongly adverse to the idea of competing on price. As a result, there was a reluctance to information customers about the costs of products. It was only due to the overall cohesion in the service matrix between customer needs (including a need for price information) and competitor positioning (showing that competitors were much better at providing price information) that this topic came up in the MuCh-QFD session (not in the FE session) and was subsequently admitted to be an important competitiveness issue. It had also to do with the presence of an experienced sales person representing the customer perspective. This illustrates the fact that a good customer advocate is needed in service definition sessions. The MuCh-QFD group was relatively self-critical. In comparison to our observations, they gave relatively low scores (bottom left figure) for channel coherence, competitor focus and progression. In the FE session the discussion on investment-readiness (in part III) was particularly useful in identifying some of the bottlenecks for e-service success. The FE group was relatively positive in its channel coherence and channel synergy scores (bottom right figure), although their scores were confirmed neither by external observations (top two figures on the right) nor our own observations.

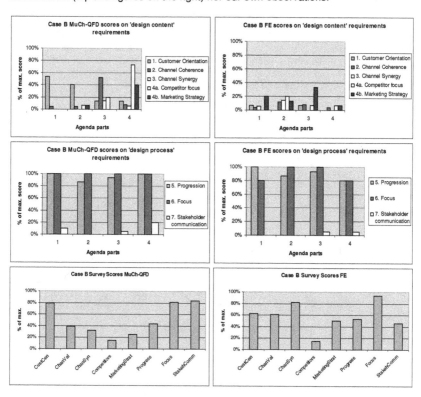

Case C: Intermediary portal

In the FE session of this case, one of the participants could be described as 'a traveler from the future'. His company had started planning the same e-service 4 years ago, and it had become commercially operational two years earlier. He provided a great deal of insight into a) what is a clever way to tackle problem XYZ?' and b) what do customers want, buy, and respond best to?' His input had a marked impact on the session. For one thing, he emphasized the importance of customer focus and the need to connect all service choices to customer priorities. As a consequence, the outputs of the FE session were as customer-oriented as the MuCh-QFD outputs, although the process was less recognizably customer-oriented. Secondly, although the first two agenda parts were still relatively democratic (caused by the GroupSystems™ approach), from part III on (a more free format breakout activity) his influence became very strong. Basically he was lecturing the others: this is what should come first and this what should come second on the site, this is how customers respond to these service functions, this is how an insurance agent can educate customers on self-service, this is how you can outperform the competition etc. His experience with regard to this e-service far exceeded that of the others, which led to a solution and strategic positioning that very much resembled his own e-service. However, the output was not generated via a group effort. This explains why the level of customer orientation and marketing strategy consideration that is found in the output is not reflected in the group process scores of participants (bottom right figure). The MuCh-QFD session was of a high quality, and it was concluded by a relatively well rounded positioning discussion. Still, compared to the other cases the output of the MuCh-QFD session was not much better, and on some points even less specific, than the FE output, due to the 'traveler from the future' in the FE session.

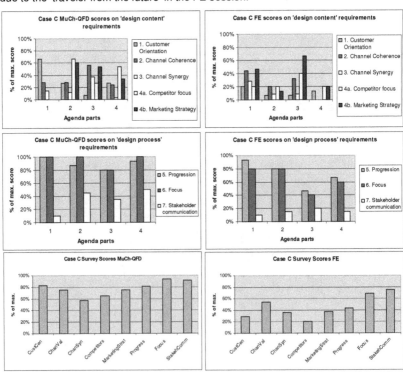

Case D: Offer registration

This case was interesting in the sense that at the start of the intake it was not focused on the end-user. It's aim was to provide a number of technical solutions to insurance agents, but initially whether and how it would fit in the commercial offer to the end-customer was not considered. However, the intake questions did what they were supposed to do in that they made the participants focus more on the end-user. In this case the participants of the FE and MuCh-QFD sessions were quite divided in their opinions as to how the e-service should be positioned and what the chances of commercial success were. (In the other cases the session outcomes were much more similar). The e-service had to compete with a paper-based solution. Although in the MuCh-QFD session the paper-based solution scored seriously lower, in the FE session it was stated that the paper-based solution would be the most attractive solution for 80% of the intermediaries. This difference was caused by the fact that the participants of the MuCh-QFD session looked at customer priorities, while the FE session focused predominantly on insurance agent priorities. In part III of the FE session the e-service option for offer registration was basically discarded. This was based on serious discussions on marketing strategy and channel cooperation (see right hand figures). The MuCh-QFD participant questionnaire scores on competitor focus and on progression were relatively low in comparison to the external observations (upper left figures) and to our session observations.

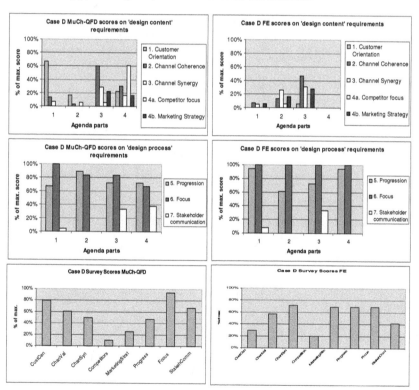

Case E: Online billing

This case was based on an e-service that was inherently very customer-oriented: the primary goal was increasing loyalty. Also, the first release was already available and the sessions focused on the second release. Hence, the participants had more prior experience with the e-service than in the other cases. Both these elements were noticeable during the sessions. Firstly, due to the loyalty objectives both sessions were relatively customer-oriented (so in that respect there was little difference between MuCh-QFD and FE sessions for this case). Secondly, the manager present in the MuCh-QFD session stated that if they had used this matrix in the first e-service release, his investment priorities with regard to Internet functions would have been different due to the customer priority insights the matrix provides. Also, he said that this approach should be used as the basis for the entire innovation process management (which is very close to the QFD philosophy). MuCh-QFD participants gave relatively low scores for competitor focus in comparison to external observations (upper left figurers) and to our own observations, maybe due to the limited availability of competitor data. The FE scores for channel coherence were relatively low in comparison to our observations and external observers: this session did involve a discussion on the relationships between online and paper-based billing, plus the role of account managers and helpdesks. Two other remarks on this case are that 1) at the end of the FE session participants had to conclude that the resulting e-service had become too complex for customers and that simplification would not be easy, and that 2) in the MuCh-QFD session participants were startled by the amount of multi-channel coordination that would be required to make this service successful (which made participants appreciate the added value of part III of the MuCh-QFD agenda).

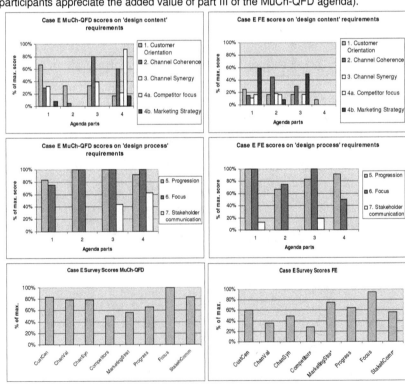

204

Case F: Telecom bundle

The e-service idea that was discussed in Case F was a relatively young one. The intake really helped progress people understand what the e-service should and should not be. The intake was also very much needed to create focus for the session. Part of the challenge was the breadth and complexity of the e-service, which encompassed promotion and selling of three quite different core products as a single product. In the FE session the group encountered particular difficulties in defining the core solutions on the website. The free format approach, in combination with the fact that nobody as yet had a very clear idea about the e-service made it difficult to make progress. In the MuCh-QFD session it was particularly the value of part III that became apparent, and participants were surprised by the multi-channel aspects that played a role in service success and they were glad this was part of the agenda. This did not emerge in the FE session. One final remark we want to make in relation to this case, and that is that the firm was reluctant to compete on price, or even on a (temporary) financial advantage for the customer. They very much liked the slogan of one of the competitors, 'combineer en profiteer' (combine & benefit), but they were very reluctant to use anything similar. They said this would not fit their brand, even though at a factual and strategic level the e-service did contain a basic price-related element. However, the corporate culture of the seemed to 'prohibit' communicating anything of the kind. We thought that was striking, especially for the MuCh-QFD session, since financial benefit was an explicit customer desire. Furthermore, MuCh-QFD participants gave relatively low scores on competitor focus, even though competitors were seriously discussed and used for comparisons.

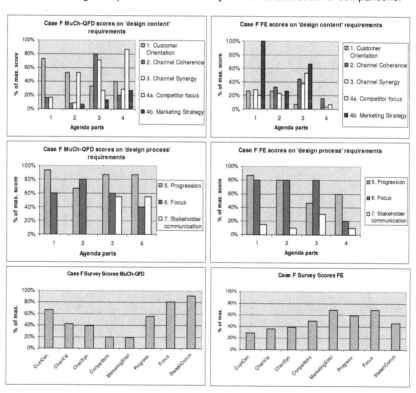

Session	Customer orientation	Channel coherence	Channel synergy	Competitive positioning
Case A MuCh-QFD	12 user needs mentioned, summarized in 6 user requirements. Focus is on customers process management and reports of financial implications for customers. Functionality: checklist and process management, (e-mail) notifications and direct links from agent Internet site to insurer.	Nine remarks made on channel coherence mainly directed towards process optimization and decision support role of agent (e-mail notifications, links to other organizations, telephone service as alternative front office, personal helpdesk).	Value chain integration is inherent in the core product. At an e-service level there are efficiencies by integrating with the agent administration. Automatically sending absence reports to accountant and insurer.	Services concepts scores better (354) than present system (180) and system of competitors (248) on user requirements
Case A FE	Customer orientation is quite central to the absence management core product. At the e-service level 7 of the 24 objectives are customer-oriented. Focus is on rapid reintegration of employee, less paper work, simplicity and lock-in relationship. Functionality: administration and history of employee absence, process management, and visualization.	Internet as backbone for other contacts. This is taken as a given and not discussed extensively: e-mail notifications; online visualization of process with options to call in personal advice.	Internet technology is used for centralized workflow management. This creates efficiencies in all channels and eliminates paper work.	No assessment made of competitive position, although it was expected that an interlocked relationship between agent and customer would emerge.
Case B MuCh-QFD	27 user needs mentioned, summarized in 6 user requirements. Focus is on information on agent, agent services and processes, tailored advice to customer, accountability. Functionality: tools for calculations of premium of insurance bundle, process information, and advice tool for specific profiles or situations.	Multi-channel issues are explicitly discussed, especially in follow-up phases: negotiation, complex questions etc. Especially in case of 'life events' (with high- en low-risk products) personal contact is expected to add value.	Value chain integration, process integration is a prerequisite for portal.	Service concept scores better (381) than present system (171) and system of direct writers (316) or independent Internet-based agents (338) on user requirements.

Case B FE	Of the 18 objectives formulated only one was related to the customer, focus was primarily on insurer and agent benefits. End-user requirements were voted least important. Focus of insurer is on increased sales, process optimization, integration of agents in value chain, and reinforcement of competitiveness of agent. Functionality: smart forms, reuse of data, process information, value chain integration.	Mild multi-channel focus. Only two items: Use and display information online from previous contacts. Agent can be asked to consult on online offers.	Existing situation is that agents have their own site. Goal is benefits of scale by centralization and chain integration. And connection to agent info systems.	It was expected that the e-service concept was going to fulfill end-user needs in better way and that the competitive position of the agent would be improved. Economies of scale for insurer, and stronger ties insurer-agent were mentioned.
Case C MuCh-QFD	28 user needs identified, summarized in 8 user requirements. Focus on Price information and attractive price quotes, price comparison, quality of product and branding. Functionality: tools for price comparison, product information, call-back functionality.	Personal face-to-face advice was considered so important that it was explicitly named as the most important functionality. In response to customer habit s, to educate them on Internet advantages, or for additional needs. Use of the Internet starting point for pro-active personalization of services.	A concern is whether enough parties will take part to generate scale advantages (which is a core assumption). Chain integration is not discussed much, but assumed as essential.	Service concept scores better (497) than present system (423) and system c direct writers (441) or bank (461) on user requirement:
Case C FE	Of the 16 objectives formulated nine were related to the customer. Meeting customer needs was considered the most important objective. Focus on accessibility, ease of use and interactivity as objective, but also price, quality ratio, and improve competitiveness of agents. Functionality defined in terms of options for price and product comparison, decision support tool, added value of insurance agent.	User needs were explicitly discussed from a coherence point of view, specifically the added value of the agent was discussed from a multi-channel perspective, but also other examples were given.	Process automation is considered to offer efficiencies both in the online and offline approach. Focus is on workflow management	Competitive position explic discussed as one of the objectives to be achieved with the e-service concept. Starting on the basis of pri competition and than the customer gets locked in du to ease of use and convenience.

Case D MuCh-QFD	20 users needs identified, grouped in six user requirements. Focus on clear costs and benefits, process overview, product comparison, product information, import in own administration and process transparency. Functionality: product information in combination with self management and support tools, insight into existing policies and conditions (by customer and by agent), and support of personal contact.	Advice is face-to-face (maybe supported by online tools). Added value from consulting product advice online, and maybe for pre-sales orientation. 'Contact me now' support is important. Additional personal contact needed when laws or personal situation change.	'Lightweight' integration into existing processes and systems is taken as a necessity. In the longer run increasing integration and efficiencies are expected.	Service concept scores better (419) than 'paper tra alternative (245) or solutior from system houses (369) on user requirements
Case D FE	Customer orientation is very secondary. Requirements are defined in terms of interest of insurance agents, discussion of customer needs are from the agent perspective: generating leads, analysis of needs for new product offerings. Focus on control of liability as a consequence of new regulation is main driver for the service concept, commercial objectives are secondary. Registration of advice is an important driver as process optimization. Functionality is mainly defined from the perspective of service provider, i.e. intermediary.	The e-service idea is inherently multi-channel. Advice seen as strictly offline; compliance with registration laws paper-based. Online option to read back product advice may be useful. Use online customer advice tools internally for lead qualification.	Minor references are made to channel synergy. Skepticism with regard to the e-service's ability to generate savings.	Commercial objectives are an important driver, not so much in terms of competition but more in terms of up- and cross-selling.
Case E MuCh-QFD	20 users needs identified, grouped in six user requirements Focus on quality and usability, ease of use, cost control and advise, personal support, negotiation and dispute settlement. Functionality: advice and one-stop shopping, analysis and reports, personalization, personal communication and export of compatibilities.	Adding depth and value to existing info and channels is central. Many (17) opportunities are addressed and discussed. Feedback that the method raises an important concern: how to manage other channels in such a way that they can provide support for customer questions?	Some call reduction is expected due to self-service. But online analyses may also raise certain questions. Less paper billing is foreseen, but not prime goal.	Service concept scores better (414) than present system (294) and 'competitor 1' (308) on use requirements. Branding and reputation ar important: company expec to offer the service before competitors do.

Case E FE	Customer orientation is central to the service. From 38 objectives 14 are directly related to customers. Customer focus is strong. Focus: customer control, customer satisfaction and loyalty, ease of use. Functionality: analysis of usage patterns, pro-active advise, payment management, self-service, information and support, one stop billing.	Implicitly the organization is assumed to be efficient and prepared to deal with multi-channel issues. Self-management by customers is seen as key.	Channel synergy is incidentally discussed, with an open eye to opportunities to reap benefits like saving on paper billing.	Basically the organization i very professional, and aware of competitors' behavior. Competitive advantages are defined in competitive terms.
Case F MuCh-QFD	20 users needs identified, grouped in six user requirements Focus on complete information on product and prices, personal advice and opportunities to compare. Functionality: face-to-face contact, tools for advice, product and price information and direct 'contact me now' opportunities	More than 14 different options for multi-channel coordination are mentioned. Strong emphasis on personal contact: for product touch and feel, for confirmation, or specific (complex) questions. Explicit feedback that method raised multi-channel awareness.	When filling Internet forms information on customers (name etc) and their subscriptions should be used to aid customers and reduce errors.	Service concept scores better (560) than present system (427) and competit (394) on user requirement:
Case F FE	Customer orientation is limited: perspective on the customer is defined more in terms of sales opportunities than based on an analysis of user needs. Only ease of use, 24/7 are incidentally mentioned. Focus: Additional sales, simple and pleasant choice process, better positioning of enterprise, and efficiency for enterprise. Functionality: show product benefits, cross and up selling, one-stop-shopping and personalization.	Few references to other channels were made.	Reuse of customer information across contacts and channels was mentioned, mostly from a value added perspective. Efficiencies were considered the least important goal.	Positioning is based on brand name of organizatio

Appendix I: Statistical details for section 7.2.4 - Disturbing factors

Interaction effect between stakeholder perspective (question 2) and session type for competitor focus

Competitor focus Anova: session type x stakeholder perspective

Dependent Variable: competitor focus

Source		Type III Sum of Squares	df	Mean Square	F	Sig.
Intercept	Hypothesis	63,164	1	63,164	21,882	,001
	Error	26,717	9,256	2,887(a)		
type2 (MuCh-QFD or FE)	Hypothesis	2,752	1	2,752	2,575	,117
	Error	37,608	35,187	1,069(b)		
v2 (stakehold. perspect.)	Hypothesis	16,263	5	3,253	**5,710**	**,007**
	Error	6,373	11,187	,570(c)		
type2 * v2	Hypothesis	1,475	4	,369	,155	,959
	Error	90,117	38	2,371(d)		

a ,647 MS(v2) + ,027 MS(type2 * v2) + ,326 MS(Error)
b ,650 MS(type2 * v2) + ,350 MS(Error)
c ,900 MS(type2 * v2) + ,100 MS(Error)
d MS(Error)

Means for competitor focus vary per perspective, mainly marketing & operations/IT

Session type	Perspective in session	Mean	N	Std. Deviation
FE	customer	1,2500	8	1,16496
	marketing	1,6667	6	1,96638
	sales	,7500	4	,95743
	operations/IT	,0000	4	,00000
	innovation	1,5000	2	,70711
	distribution	2,0000	1	.
	Total	1,1200	25	1,30128
MuCh-QFD	customer	1,5000	6	1,87083
	marketing	2,6667	6	2,25093
	sales	1,6000	5	1,51658
	operations/IT	1,0000	6	1,09545
	distribution	2,0000	1	.
	Total	1,7083	24	1,70623
Total	customer	1,3571	14	1,44686
	marketing	**2,1667**	**12**	**2,08167**
	sales	1,2222	9	1,30171
	operations/IT	**,6000**	**10**	**,96609**
	innovation	1,5000	2	,70711
	distribution	2,0000	2	,00000
	Total	**1,4082**	**49**	**1,52641**

Marketing perspective significantly different?

	marketing	N	Mean	Std. Deviation	Std. Error Mean
so_obj_4 aa competit or focus	yes	12	2,1667	2,08167	,60093
	no	37	1,1622	1,23634	,20325

t-test 'marketing perspective' versus other stakeholder perspectives

	Levene's Test for Equality of Variances		t-test for Equality of Means		
	F	Sig.	t	df	Sig. (2-tailed)
Equal variances assumed	13,124	,001	2,046	47	,046
Equal variances not assumed			1,583	13,606	,136

Operations/IT perspective significantly different?

	operat_IT	N	Mean	Std. Deviation	Std. Error Mean
so_obj_4aa competitor focus	yes	10	,6000	,96609	,30551
	no	39	1,6154	1,58306	,25349

t-test 'operations/IT perspective' versus other stakeholder perspectives

	Levene's Test for Equality of Variances		t-test for Equality of Means		
	F	Sig.	t	df	Sig. (2-tailed)
Equal variances assumed	2,185	,146	-1,929	47	,060
Equal variances not assumed			-2,558	23,069	,018

Interaction effect between previous session experience (question 3) and session type for channel synergy

Channel synergy Anova: session type x previous session experience

Dependent Variable: channel synergy

gt		Type III Sum of Squares	df	Mean Square	F	Sig.
Intercept	Hypothesis	733,398	1	733,398	4807,781	,009
	Error	,153	1	,153(a)		
type2 (MuCh-QFD or FE)	Hypothesis	,273	1	,273	,012	,930
	Error	22,554	1	22,554(b)		
v3 (previous sessions)	Hypothesis	,153	1	,153	,007	,948
	Error	22,554	1	22,554(b)		
type2 * v3	Hypothesis	22,554	1	22,554	**6,711**	**,013**
	Error	151,239	45	3,361(c)		

a MS(v3)
b MS(type2 * v3)
c MS(Error)

	MuCh-QFD	FE
Session experience	3.3 (n=12)	4.5 (n=11)
No session experience	4.6 (n=12)	3.1 (n=14)

Interaction effect between satisfaction (factor of questions 5a to 5c) and session type for channel synergy

Channel synergy Anova: session type x satisfaction

Dependent Variable: channel synergy

Source		Type III Sum of Squares	df	Mean Square	F	Sig.
Intercept	Hypothesis	288,989	1	288,989	56,899	,000
	Error	64,211	12,643	5,079(a)		
type2 (MuCh-QFD or FE)	Hypothesis	8,098	1	8,098	1,265	,308
	Error	34,422	5,376	6,403(b)		
v40 (satisfact)	Hypothesis	43,080	7	6,154	,845	,600
	Error	32,733	4,492	7,287(c)		
type2 * v40 (session * satisfact)	Hypothesis	32,209	4	8,052	**2,857**	**,038**
	Error	98,631	35	2,818(d)		

a ,605 MS(v40) + ,047 MS(type2 * v40) + ,349 MS(Error)
b ,685 MS(type2 * v40) + ,315 MS(Error)
c ,854 MS(type2 * v40) + ,146 MS(Error)
d MS(Error)

Linear effect between satisfaction (factor of questions 5a to 5c) and focus

Focus Anova: session type x satisfaction

Dependent Variable: focus

Source		Type III Sum of Squares	df	Mean Square	F	Sig.
Intercept	Hypothesis	277,061	1	277,061	277,314	,000
	Error	12,211	12,222	,999(a)		
type2 (MuCh-QFD or FE)	Hypothesis	,154	1	,154	,428	,520
	Error	7,269	20,158	,361(b)		
v40 (satisfact.)	Hypothesis	8,664	7	1,238	**4,435**	**,019**
	Error	2,648	9,487	,279(c)		
type2 * v40	Hypothesis	,834	4	,208	,302	,875
	Error	24,193	35	,691(d)		

a ,605 MS(v40) + ,047 MS(type2 * v40) + ,349 MS(Error)
b ,685 MS(type2 * v40) + ,315 MS(Error)
c ,854 MS(type2 * v40) + ,146 MS(Error)
d MS(Error)

Pearson correlation satisfaction x focus

		v40
focus	Pearson Correlation	,400(**)
	Sig. (2-tailed)	,005
	N	48

** Correlation is significant at the 0.01 level (2-tailed).

Linear effect between participation during intake and channel coherence

Channel coherence Anova: session type x participation during intake

Dependent Variable: channel coherence

Source		Type III Sum of Squares	df	Mean Square	F	Sig.
Intercept	Hypothesis	374,649	1	374,649	24,872	,126
	Error	15,063	1	15,063(a)		
type2 (MuCh-QFD or FE)	Hypothesis	6,819	1	6,819	2237,290	,013
	Error	,003	1	,003(b)		
v34 (intake participat.)	Hypothesis	15,063	1	15,063	**4942,090**	**,009**
	Error	,003	1	,003(b)		
type2 * v34	Hypothesis	,003	1	,003	,001	,977
	Error	158,082	45	3,513(c)		

a MS(v34)
b MS(type2 * v34)
c MS(Error)

	MuCh-QFD	FE
Intake participation	5.7 (n=6)	4.5 (n=5)
No intake participation	3.9 (n=18)	2.8 (n=20)

Interaction effect between timing (morning/afternoon) and session type for customer orientation

Customer orientation Anova: session type x timing (morning vs. afternoon)

Dependent Variable: customer orientation

Source		Type III Sum of Squares	df	Mean Square	F	Sig.
Intercept	Hypothesis	668,889	1	668,889	177,460	,048
	Error	3,769	1	3,769(a)		
type2 (MuCh-QFD or FE)	Hypothesis	35,558	1	35,558	5,402	,259
	Error	6,582	1	6,582(b)		
v38 (morning/ afternoon)	Hypothesis	3,769	1	3,769	,573	,588
	Error	6,582	1	6,582(b)		
type2 * v38 (session * timing)	Hypothesis	6,582	1	6,582	**5,313**	**,026**
	Error	55,750	45	1,239(c)		

a MS(v38)
b MS(type2 * v38)
c MS(Error)

	MuCh-QFD	FE
Morning session	4.7 (n=16)	3.7 (n=9)
Afternoon session	4.9 (n=8)	2.3 (n=16)

Linear effects for intention to use session results- Effects were analyzed for FE and MuCh-QFD groups separately.

FE correlations intention to use with channel synergy, marketing strategy, progression and focus

		Intention to use results?
channel synergy	Pearson Correlation	0.367
	Sig. (2-tailed)	0.071
	N	25
marketing strategy	Pearson Correlation	**0.397(*)**
	Sig. (2-tailed)	**0.050**
	N	25
progression	Pearson Correlation	0.387
	Sig. (2-tailed)	0.056
	N	25
focus	Pearson Correlation	**0.645(**)**
	Sig. (2-tailed)	**0.000**
	N	25

** Correlation is significant at the 0.01 level (2-tailed).
* Correlation is significant at the 0.05 level (2-tailed).

MuCh-QFD correlations intention to use with channel synergy, marketing strategy, progression and focus

		Intention to use results?
channel synergy	Pearson Correlation	,611(**)
	Sig. (2-tailed)	,002
	N	24
marketing strategy	Pearson Correlation	,420(*)
	Sig. (2-tailed)	,041
	N	24
progression	Pearson Correlation	,622(**)
	Sig. (2-tailed)	,001
	N	24
focus	Pearson Correlation	,494(*)
	Sig. (2-tailed)	,014
	N	24

* Correlation is significant at the 0.05 level (2-tailed).
** Correlation is significant at the 0.01 level (2-tailed).

Appendix J: Abbreviations

B2B	Business to Business
B2C	Business to Consumer
BB	BroadBand
CxO	Chief xx Officer (e.g. Chief Executive, Information or Technology Officer)
FE	Fundamental Engineering
GDSS	Group Decision Support Systems
HoQ	House of Quality
ICT	Information and Communication Technology
IPA	Importance Performance Analysis
ISI	Innovation Scenarios for Intermediaries
IT	Information Technology
KPN	Koninklijke PTT Nederland (Royal PTT Netherlands)
MuCh-QFD	Multi-Channel-QFD
PLACE	Physical presence and Location Aspects in electronic Commerce Environments
QFD	Quality Function Deployment
SME	Small and Medium Enterprises

References

Aken, J. E. v. (2004). "Management Research Based on the Paradigm of the Design Sciences: The Quest for Field-Tested and Grounded Technological Rules." *Journal of Management Studies* **41** (2): pp. 219-146.

Aken, J. E. v. (2005). "Management Research as Design Science: Articulating the Research Products of Mode 2 Knowledge Production in Management." *British Journal of Management* **16** (1): pp. 19-36.

Alter, S. (1999). *Information systems; A management perspective*. New York, Addison-Wesley.

Anton, J. (1996). *Call Center Management by the Numbers*. Santa Maria, CA, Press on Regardless.

Anton, J. (2000). "The past, present and future of customer access centers." *International Journal of Service Industry Management* **11** (2): 120-130.

Barnes, S. J. and R. Vidgen (2001). "An evaluation of cyber-bookshops: The WebQual method." *International Journal of Electronic Commerce* **6** (1): pp. 11-30.

Beyer, H. and K. Holtzblatt (1998). *Contextual Design: Defining Customer-Centered Systems*. San Francisco, Morgan Kaufmann Publishers.

Bhattacherjee, A. (2001). "An empricial analysis of the antecedents of electronic commerce service continuance." *Decision Support Systems* **32**: pp. 201-214.

Bouwman, H. and L. v. d. Wijngaert (2003). *E-commerce B2C Research in Context: Policy Capturing and Customer Value*. Proceedings of the 16Th Bled eCommerce Conference eTransformation.

Bowersox, D. J. and M. Bixby-Cooper (1992). *Strategic Marketing Channel Management*. New York, McGraw-Hill.

Bucklin, L. P. (1966). *A Theory of Distribution Channel Structure*. Berkeley, CA, IBER special publications.

Bucklin, L. P. (1972). *Competition and Evolution in the Distributive Trades*. Englewood Cliffs, NJ, Prentice Hall.

Buede, D. M. (2000). *The Engineering Design of Systems: Models and Methods*. New York, John Wiley & Sons.

Campbell, D. T. (1957). "Factors relevant to the validity of experiments in social settings." *Psychological Bulletin* **54**: pp. 297-312.

Cassidy, J. (2002). *Dot.con: The Greatest Story Ever Sold*. New York, Harper Collins.

Chan, L.-K. and M.-L. Wu (2002). "Quality function deployment: A literature review." *European Journal of Operational Research* **143**: 463-497.

Change-Sciences-Group (2004). *Online Brokerage Customer Acquisition; Customer Experience Benchmark Report Overview*. Change Sciences Group, Accessed www.changesciences.com/pub/Brokerage/Acquisition/CEBR_OnlineBroker ageCustomerAcquisition_Q1_2004_Overview.pdf.

Chen, L., M. L. Gillenson and D. L. Sherrell (2002). "Enticing online customers: an extended technology acceptance perspective." *Information & Management* **39**: pp. 705-719.

Clark, G., R. Johnston and M. Shulver (2000). Exploiting the service concept for service design and development. *New Service Design*. J. Fitzsimmons and M. Fitzsimmons. Thousand Oaks, CA, Sage: pp. 71-91.

Clausing, D. P. (1994). *Total Quality Development: A Step-by-Step Guide to World-Class Concurrent Engineering*. New York, ASME Press.

Cohen, L. (1995). *Quality Function Deployment: how to make QFD work for you*. Reading, Mass., Addison-Wesley.

Cross, N. (1994). *Engineering Design Methods; Strategies for Product Design*. Chichester, John Wiley & Sons.

DeLone, W. H. and E. R. McLean (2003). "The DeLone and McLean Model of Information Systems Success: A Ten-Year Update." *Journal of Management Information Systems* **19** (4): pp. 9-30.

Dilthey, W. (1900). Die Entstehung der Hermeneutik. *In: Ges. Werke, Bd. 5. (1957)*. Stuttgart/Göttingen.

Dobler, D. W. and D. N. Burt (1996). *Purchasing and Supply Management*. New York, McGraw-Hill.

Dukcevich, D. (2004). *Ameritrade Vs. E*Trade*. Forbes, Accessed http://www.forbes.com/2004/01/12/cx_dd_0112mondaymatchup.html.

Evans, P. and T. S. Wurster (1999). "Getting Real About Virtual Commerce." *Harvard Business Review* **77** (Nov-Dec): pp. 85-94.

Evans, P. and T. S. Wurster (2000). *Blown to Bits: How the Economics of Information transforms Strategy*. Boston, Mass., HBSP.

Gadamer, H. G. (1960). *Wahrheit und Methode; Grundzuge einer philosophischen Hermeneutik*. Tubingen, Mohr.

Gates, B. (1995). *The Road Ahead*. New York, Penguin Books.

Gebauer, J. and A. Scharl (1999). "Between flexibility and automation: An evaluation of Web technology from a business process perspective." *Journal of Computer- Mediated Communication* **5** (2): Retrieved July 2002 from http://www.ascusc.org/jcmc/vol5/issue2.

Geertz, C. (1973). *The Interpretation of Culture*. New York, Basic Books.

Gilder, G. (1994). *Life After Television: The Coming Transformation of American Life*. New York, W. W. Norton Company.

Goldstein, S. M., R. Johnston, J. Duffy and J. Rao (2002). "The service concept: the missing link in service design research?" *Journal of Operations Management* **20** (2): pp. 121-134.

Gordijn, J. (2002). *Value-based Requirements Engineering; Exploring Innovative eCommerce Ideas. Faculty of Exact Sciences*. Amsterdam, Vrije Universiteit: 292.

Grönroos, C. (1993). "Toward a third phase in service quality research: challenges and future directions." *Advances in Services Marketing and Management* **2**: pp. 49-64.

Grönroos, C. (1994). "From marketing mix to relationship marketing: Towards a paradigm shift in marketing." *Management Decision* **32** (2):pp. 4-20.

Grönroos, C. (2000). *Service Management and Marketing: A Customer Relationship Management Approach*. New York, John Wiley & Sons.

Grönroos, C., F. Heinonen, K. Isoniemi and M. Lindholm (2000). "The NetOffer model: a case example from the virtual marketspace." *Management Decision* **38** (4): pp. 243-252.

Gummesson, E. (1993). *Quality Management in Service Organizations*. Stockholm, Stockholm University.

Hagenaars, J. A. P. and J. H. G. Segers (1980). Onderzoeksontwerp (Research Design). In: *Sociologische Onderzoeksmethoden. Deel II; Technieken van Causale Analyse (Sociological Research Methods, Volume II: Techniques of Causal Analysis)*. J. H. G. Segers and J. A. P. Hagenaars (ed.). Assen, Van Gorcum. **Volume II**: 369.

Harink, J. H. A. (1997). *Excelleren met Electronisch Inkopen; De Snelweg naar betere Prestaties*. Deventer, Kluwer Bedrijfsinformatie.

Heijden, H. v. d. and P. Valiente (2002). *The value of mobility for business process performance: evidence from Sweden and the Netherlands*. 10th European Conference on Information Systems, Gdansk, Poland.

Hennig-Thurau, T. and U. Hansen (2000). Relationship Marketing; Some reflections on the State-of-the-Art of the relational concept. *Relationship Marketing; Gaining competitive advantage through customer satisfaction and customer retention.* T. Hennig-Thurau and U. Hansen. Berlin, Springer-Verlag: pp. 3-27.

Herzwurm, G., S. Schockert, U. Dowie and M. Breidung (2002). *Requirements engineering for mobile-commerce applications*, Athens, Greece.

Heskett, J. L., W. E. Sasser and L. A. Schlesinger (1997). *The Service Profit Chain: How leading companies link profit and growth to loyalty, service and value.* New York, Free Press.

Hevner, A. R., S. T. March, J. Park and S. Ram (2004). "Design Science in Information Systems Research." *MISQ* **28** (1): pp. 75-105.

Hill, N. and J. Alexander (2000). *Handbook of Customer Satisfaction and Loyalty Measurement.* Hampshire, Gower.

Holsti, O. R. (1969). *Content Analysis for the Social Sciences and Humanities.* Reading, Mass., Addison-Wesley.

Hull, E., K. Jackson and J. Dick (2002). *Requirements Engineering; a structured project information approach.* London, Springer-Verlag.

Jiang, J. J., G. Klein and C. L. Carr (2002). "Measuring Information Systems Service Quality: SERVQUAL from the Other Side." *MIS Quarterly* **26** (2): pp. 145-166.

Johnston, R. (1999). "Service operations management: Return to roots." *International Journal of Operations & Production Management* **19** (2): pp. 104-124.

Kar, E. A. M. v. d. (2004). *Designing Mobile Information Services; An approach for Organisations in a Value Netwerk. Technology, Policy and Management.* Delft, Delft University of Technology.

Keen, P. G. W. and C. Ballance (1997). *On-line Profits, A Manager's Guide to Electronic Commerce.* Boston, Mass., HBSP.

Kettinger, W. J. and C. C. Lee (1994). "Perceived Service Quality and User Satisfaction with the Information Services Function." *Decision Sciences* **25** (5/6): pp. 737-765.

Kotler, N. (1999). *Kotler on Marketing : How to Create, Win, and Dominate Markets.* New York, Free Press.

Lynn, F. (2000). The dynamics and economics of channel marketing systems. *The Handbook of Business Strategy.* NY, Faulkner & Gray.

Malone, T., J. Yates and R. Benjamin (1987). "Electronic markets and electronic hierarchies: Effects of information technology on market structure and corporate strategies." *Communications of the ACM* **30** (6): pp. 484-497.

Martilla, J. and J. James (1977). "Importance-performance analysis." *Journal of Marketing* **41** (January): pp. 77-79.

Mazur, G.H. (1993). *QFD for Service Industries; From Voice of Customer to Task Deployment.* 5th Symposium on Quality Function Deployment, Novi, Michigan.

Mazur, G.H. (1995). *QFD Applications in Health Care and Quality of Work Life.* First International Symposium on QFD, Tokyo, Japan.

Menor, L. J., M. V. Tatikonda and S. E. Sampson (2002). "New service development: areas for exploitation and exploration." *Journal of Operations Management* **20** (2): pp. 135-157.

Mizuno, S. and Akao, Y. ed. (1994). *Qfd: The Customer-Driven Approach to Quality Planning & Deployment.* Tokyo, Japan, Asian Productivity Organization.

Modahl, M. (2000). *Nu of Nooit: Wat bedrijven moeten doen om de internetconsument voor zich te winnen.* Schoonhoven, Academic Service.

Negroponte, N. (1995). *Being Digital.* New York, Alfred A. Knopf.

Nielsen, J. (2004a). *Card Sorting: How many users to test.* Nielsen Norman Group, Accessed 15-4-2005, http://www.useit.com/alertbox/20040719.html.

Nielsen, J. (2004b). *Why you only need to test with 5 users.* Useit.com Alertbox, Nielsen Norman Group, Accessed 2005, http://www.useit.com/alertbox/20000319.html.

Nielsen, J. and T. K. Landauer (1993). *A mathematical model of the finding of usability problems.* Proceedings ACM INTERCHI'93 Conference, Amsterdam.

Norman, D. A. (2004). *Emotional Design: Why We Love (or Hate) Everyday Things.* New York, Basic Books.

Normann, R. A. (2000). *Service Management; Strategy and Leadership in Service Business.* Chichester, J. Wiley & Sons.

Parasuraman, A., L. L. Berry and V. A. Zeithaml (1985). "A conceptual model of service quality and its implications for future research." *Journal of Marketing* **49** (4): pp. 41-50.

Parasuraman, A., L. L. Berry and V. A. Zeithaml (1993). "More on improving the measurement of service quality." *Journal of Retailing* **96** (1): pp. 140-147.

Pitt, L. F., R. T. Watson and C. B. Kavan (1995). "Service Quality: A Measure of Information Systems Effectiveness." *MIS Quarterly* **19** (2): pp. 173-188.

Planet-Internet (2004). *Zoekmachines tonen Nederlanders de weg.* Planet Internet Multimedia Nieuws, Accessed http://planet.nl/planet/show/id=118880/contentid=434196/sc=d2e3da.

Porter, M. (1999). *Porter over concurrentie*, Business contact.

Porter, M. E. (2001). "Strategy and the Internet." *Harvard Business Review* **79** (3): pp. 63 - 78.

Power, A. (2000). "Channel surfing." *Outlook* (1): pp. 54-61.

Prahalad, C. K. (1998). "Managing discontinuities: The emerging challenges." *Research-Technology Management* (May-June): pp. 14-22.

Prasad, B. (1996). *Concurrent Engineering Fundamentals: Integrated Product and Process Organization.* Upper Saddle River, NJ, Prentice-Hall.

Preissl, B., H. Bouwman and C. Steinfield, Eds. (2004). *E-Life after the Dot Com Bust.* Heidelberg, Physica-Verlag.

Ramaswamy, R. (1996). *Design and Management of Service Processes.* Reading, Mass., Addison-Wesley.

Rigby, D. K., F. F. Reichheld and P. Schefter (2002). "Avoid the four perils of CRM." *Harvard Business Review* (Feb): pp. 101-109.

Schueler, R. (2003). *Channel preferences onderzoek.* Amsterdam, Interview-NSS.

Shapiro, C. and H. R. Varian (1998). *Information Rules : A Strategic Guide to the Network Economy.* Boston, Mass, HBSP.

Shostack, L. G. (1984). "Designing services that deliver." *Harvard Business Review* **62** (1): pp. 133-139.

Simons, L. P. A. (2001). *Cisco case study: Lessons for increasing the value of Web and call center channels for customer contact.* 2nd World Congress on the Management of Electronic Commerce, Hamilton, Canada.

Simons, L. P. A. and H. Bouwman (2003a). *Designing a marketing channel mix.* The 4th IBM eBusiness Conference, Guildford UK, School of Management, University of Surrey.

Simons, L. P. A. and H. Bouwman (2003b). *Developing a Multi Channel Service Model and Design Method.* 4th World Congress on the Management of Electronic Commerce, Hamilton, Canada.

Simons, L. P. A. and H. Bouwman (2004). "Designing a channel mix." *International Journal of Internet Marketing and Advertising* **1** (3): pp. 229-250.

Simons, L. P. A. and H. Bouwman (2005). "Multi-Channel Service Design Process: Challenges and Solutions." *International Journal of Electronic Business* **3** (1): pp. 50-67.

Simons, L. P. A. and H. Bouwman (2006a). "Designing a marketing channel mix." *To appear in: International Journal of IT Management* **5** (4).

Simons, L. P. A. and H. Bouwman (2006b). "Extended QFD: Multi-channel service concept design." *To appear in: Total Quality Management and Business Excellence.*

Simons, L. P. A., H. Bouwman and C. Steinfield (2002). "Strategic Positioning of the Web in a Multi Channel Market Approach." *Internet Research: Electronic Networking Applications and Policy* **12** (4): pp. 339-347.

Slaughter, N. (2004). *Reinventing Schwab.* The Motley Fool, Broker Center, Accessed http://www.fool.com/news/commentary/2004/commentary040603ns.htm.

Smits, G. J. (2006). *Beter multi-channelen.* JungleRating, Accessed April 2006, http://www.junglerating.nl/?actueel/nieuws/view/120.

Steinfield, C., H. Bouwman and T. Adelaar (2002). "The dynamics of click-and-mortar electronic commerce: Opportunities and management strategies." *International Journal of Electronic Commerce* **7** (1): pp. 93-119.

Steinfield, C., D. Wit, T. Adelaar, A. Bruins, E. Fielt, A. Hoefsloot, A. Smit and H. Bouwman (2000). *Leveraging physical and virtual presence in electronic commerce: The role of hybrid approaches in the new economy.* International Telecommunications Society, Buenos Aires, Argentinia.

Stern, L. W., A. I. El-Ansary and A. T. Coughlan (1996). *Marketing Channels.* Englewood Cliffs, NJ, Prentice Hall.

Tapscott, D. (1996). *The Digital Economy: Promise and Peril in the Age of Networked Intelligence.* New York, McGraw-Hill.

Tapscott, D. (2001). "Rethinking strategy in a networked world (or why Michael Porter is wrong about the Internet)." *Strategy + Business* (Third Quarter): pp. 34-41.

Tapscott, D., D. Ticoll and A. Lowy (2000). *Digital capital -harnessing the power of business webs*. Boston, Massachusetts, Harvard business school press.

Tax, S. S. and F. I. Stuart (1997). "Designing and implementing new services: The challenges of integrating service systems." *Journal of Retailing* **73** (1): pp. 105-134.

Tullis, T. and L. Wood (2004). *How Many Users Are Enough for a Card-Sorting Study?* Proceedings UPA'2004, Minneapolis, MN.

Venetis, K. A. (1997). *Service Quality and Customer Loyalty in Professional Business Service Relationships; An Empirical Investigation into the Customer-Based Service Quality Concept in the Dutch Advertising Industry*. Maastricht, University of Maastricht: 223.

Venkatesh, A. (1999). *Virtual models of marketing and consumer behavior*. ESRC Virtual Society Program Workshop: E-Commerce and the Restructuring of Consumption, London.

Verhagen, T., W. d. Vries and E. v. d. Ham (2001). "Succesvol online verkopen vereist inzicht in de ondersteuning van het aankoopproces." *Holland Management Review* **81**: pp. 38-49.

Vredenburg, K., S. Isensee and C. Righi (2001). *User-Centered Design: An Integrated Approach*. New Jersey, Prentice Hall.

Wang, Y., H. Po Lo, R. Chi and Y. Yang (2004). "An integrated framework for customer value and customer-relationship-management performance: a customer-based perspective from China." *Managing Service Quality* **14** (2/3): pp. 169-182.

Wigand, R. T. and R. I. Benjamin (1995). "Electronic commerce: Effects on electronic markets." *Journal of Computer Mediated Communication* **1** (3): Retrieved Jan 2002 from http://www.ascusc.org/jcmc/vol1/issue3.

Wilson, C. P. (2004). *The US Consumer 2004: Multichannel and In-Store Technology*, Forrester Research.

Wootten, G. (2003). "Channel conflict and high involvement Internet purchases - a qualitative cross cultural perspective of policing parallel importing." *Qualitative Market Research* **6** (1): pp. 38-47.

Yin, R. K. (1994). *Case Study Research*. London, Sage Publications.

Wissenschaftlicher Buchverlag bietet

kostenfreie

Publikation

von

wissenschaftlichen Arbeiten

Diplomarbeiten, Magisterarbeiten, Master und Bachelor Theses
sowie Dissertationen, Habilitationen und wissenschaftliche Monographien

Sie verfügen über eine wissenschaftliche Abschlußarbeit zu aktuellen oder zeitlosen
Fragestellungen, die hohen inhaltlichen und formalen Ansprüchen genügt,
und haben **Interesse an einer honorarvergüteten Publikation**?

Dann senden Sie bitte erste Informationen über Ihre Arbeit per Email
an info@vdm-verlag.de. Unser Außenlektorat meldet sich umgehend bei Ihnen.

VDM Verlag Dr. Müller Aktiengesellschaft & Co. KG
Dudweiler Landstraße 125a
D - 66123 Saarbrücken

www.vdm-verlag.de

www.ingramcontent.com/pod-product-compliance
Lightning Source LLC
LaVergne TN
LVHW022309060326
832902LV00020B/3362